BRITAIN'S MAMMALS

A field guide to the mammals of Great Britain and Ireland

Dominic Couzens, Andy Swash, Robert Still and Jon Dunn

Updated Edition

WILD Guides

PRINCETON
press.princeton.edu

Published by Princeton University Press,
41 William Street, Princeton, New Jersey 08540
In the United Kingdom: Princeton University Press, 6 Oxford Street,
Woodstock, Oxfordshire OX20 1TR
press.princeton.edu

Requests for permission to reproduce material from this work should be sent to
permissions@press.princeton.edu

First published 2017
Updated edition 2021

British Library Cataloging-in-Publication Data is available

Library of Congress Control Number 2021935496
ISBN 978-0-691-22471-8
Ebook ISBN 978-0-691-22486-2

Production and design by **WILD**Guides Ltd., Old Basing, Hampshire UK.
Printed in Italy

10 9 8 7 6 5 4 3 2 1

Contents

Foreword

Blessed with a highly advanced central nervous system, astute senses and with wits to match, mammals are tricky to see and identify, and mammal watching has not perhaps caught on in the same way as birdwatching. This does, however, seem a bit strange given their obvious appeal. A magazine with one of their fluffy faces set with a soulful and watery gaze will sell better than anything. It's almost as if we see a connection in those eyes – we are, after all, genetically closer to, say, a Badger than we are a Blue Tit. We are, after all, mammals too.

From an early age we're told stories that feature foxes, badgers, mice and moles: we're pre-conditioned to have some kind of affinity with mammals. Yet given their highly strung nature, the fact that the majority of them are prey to most predators, or are hunted or persecuted by us, apart from the odd squirrel, rat or fleeting glimpse of something furry scarpering across the road, we rarely get to see any of them – they become almost abstract to our everyday lives, merely symbolic representations.

Even if you spend all your time out of doors, seeing any of our 122 different species of mammal, other than a Grey Squirrel, is a rare and special thing. Partly because of this rarity and exclusivity, mammals are exciting – of that there is no doubt.

Mammals, even if we don't see them most of the time, have a hold on us. With many of us becoming increasingly disconnected with the natural world (and that includes townies, country folk and even farmers), these species slip further and further from our everyday consciousness.

This is why this book, written by, and for that matter produced by, some very clever mammals, is such a milestone. We've not had a field guide like this on this subject for some time – sure there have been huge, detailed, systematic and specialist accounts, but the field guides have been a re-hash of a book that I had as a child. Once again, **WILD**_Guides_ have turned out a top-notch guide in their own inimitable style – which means it's up-to-date, delicious to look at and exceptionally user-friendly. This means that for every one of those brief moments you might have with a British or Irish mammal you stand a good chance of working out which one of them it was. And if you want to go a step or two further and improve the odds of seeing one of the fluffies that share our landscape there are plenty of tips to make you a better mammal-watcher too.

Nick Baker

Introduction

This book is a guide to finding and identifying the mammals of Great Britain and Ireland. Its purpose is twofold. Firstly, it presents profiles on every mammal you can encounter in Britain and Ireland – including all the bats, seals and cetaceans recorded up to the end of March 2021, as well as established introduced or reintroduced species, and those exotic species that are known to have bred in the wild within the last fifty years. As well as providing details on identification, the species accounts also present information on sounds, signs, habitat, food, habits, breeding behaviour and population and status – so if you find a mammal anywhere in Britain and Ireland you will have the means to identify it and to find out more about its biology and ecology.

The second purpose of the book is to encourage people to go and look for mammals. At present, the number of mammal-watchers is far fewer than the number of bird-watchers, who must number hundreds of thousands. Although mammal-watching is still something of a niche activity, it is a growing one, as people are increasingly discovering the fascination of mammals and are keen to see them more often. If this book helps, even in a small way, to inspire more people to go mammal-watching, and ideally to contribute information on their distribution or biology, it will have been worth writing.

The Greater White-toothed Shrew is one of the most recent additions to the list of British and Irish mammals, being recorded in Ireland for the first time in 2007, when skulls were found in the regurgitated pellets of Barn Owls and Kestrels.

The biology and life-cycle of mammals

Mammals exhibit an extreme range of body forms: for example, the largest, the Blue Whale, is ten million times heavier than the smallest bat, and lives in the sea, while the bat flies in the air. But all mammals share certain features that define them as the taxonomic class Mammalia (see *page 16*). They all have hair (albeit sparse on many marine species), they all suckle their young on milk produced from the mammary glands, and they all have a unique arrangement of jaw bones that has given rise to the three bones of the inner ear – the malleus, incus and stapes.

Hair in the form of fur is the most obvious mammalian feature, just as feathers characterise birds. Both body coverings have powerful insulating properties and enable both groups to regulate their own body temperatures, rather than being dependent upon external heat from the sun. An internal heat source coupled with thermal regulation allows mammals to live in both warm and cold climates, and permits many to be nocturnal. The maintained body temperature of mammals varies, from 35°C in marsupials and hedgehogs to 39°C in rabbits and cats, ideal for the chemical reactions that drive living processes. Mammals are therefore often fast-moving and alert, regardless of external conditions.

The cost of maintaining a high body temperature can be considerable. Mammals use a lot of energy that must be constantly replenished, requiring an animal to find enough food to keep it alive – a shrew, its small body size making it vulnerable to heat loss, must eat every hour or so in order to survive. In places with strong seasonal variations in temperature and food supply, such as Britain, some mammals (bats, dormice and Hedgehog) pass the lean months in a state of reduced body temperature and function (torpor), known as hibernation, to save energy. Bats and dormice may also enter into a temporary torpor on cold nights outside the hibernation period, and many other species curtail their activity in winter to reduce their energy demands. Perhaps surprisingly, given their high metabolic rate and energy requirements, shrews do not hibernate. The adults die in the autumn, and only young animals survive the winter, albeit with high mortality; this is balanced by a high reproductive output in the summer months.

Mammals suckle their young with milk from their mammary glands, but this is only one distinctive feature of mammal reproduction. With a tiny number of exceptions (egg-laying mammals or Monotremes), the fertilized embryos develop inside the mother's uterus (an enlarged cavity in the oviduct) and are born live. Fertilization is also internal. The length of gestation varies between the two main divisions of mammals. In marsupials, represented in Britain only by the introduced Red-necked Wallaby, the embryo develops for just 12–28 days inside the womb before leaving the mother's body as a poorly-developed infant, just a few centimetres long. It is born into an external pouch, the marsupium, and using its strong forelimbs it climbs the mother's fur up to the teats. In other mammals the embryo remains inside the womb until a much later stage of development, and usually for a much longer duration than a marsupial of comparable size. The embryo is nourished by the placenta, an organ that allows the transfer of nutrients and waste products between mother and growing embryo. The gestation period of British placental mammals ranges from around 20 days in mice and rats to 18 months in the Killer Whale.

It is necessary for mammals to time their breeding, particularly the birth of young, to coincide with a time when food availability is at its peak. The breeding season – even in moles – is closely correlated with day length, but in some rodents, for example, the onset of the female fertile period (oestrus) is triggered by chemicals in sprouting spring vegetation. However, in some species, the timing of reproduction must be altered. Badgers, for instance, give birth in February; their gestation of eight weeks would mean mating in December, when they are trying to preserve body condition just to survive. Consequently, Badgers will mate at almost any time of year, but the implantation of the fertilized embryo can be delayed until the most suitable time for gestation to begin. Several other British mammals also exhibit delayed implantation, including European Roe Deer and Common Seal. There is a different timing strategy in bats: typically they copulate in the autumn, but the sperm is stored inside the female and fertilization is delayed until the natural gestation period produces births in mid-summer.

Such delay mechanisms underline the fact that many mammals do not have close, long-term parental relationships – humans are unusual in this respect. In a wide range of mammals, from mice to dolphins and deer to otters, the female is exclusively responsible for parental care, males playing no part at all. Often several males contribute to genetic makeup of a litter. For their part, males frequently compete vigorously for access to females, as exemplified by the fierce clashes over harems seen in Red Deer. Sexual selection produces differences in body shape and size, and status signals such as antlers and horns.

In some placental mammals, such as deer, seals and whales, the young are born in an advanced state of development and are capable of running or swimming soon after birth. Such precocial young start life very differently from the altricial young of many rodents, bats and carnivores, for example, which are born in a more-or-less helpless state, often naked and with the eyes closed; many remain in a nest or burrow for some time before emerging. Regardless of the gestation period or developmental state at birth, all young mammals feed on the mother's milk for a period after they are born. The process of producing milk to nourish the young is known as lactation; the young feed from the teats. Suckling may be a lengthy process, the young being weaned gradually onto solid food.

Milk-producing mammary glands are not the only glands unique to mammals: others include sweat glands (for temperature regulation) and sebaceous glands, which among other functions produce odours. Most mammals live in a much more smell-dominated world than we do, and the extrusions from these glands form an important component in communication. The glands may produce individual scents that can be deposited, along with urine and/or solid faeces, to mark territorial boundaries and to attract the opposite sex. Mammals can thus acquire much information about their neighbours without actually meeting them.

Alongside smell, mammals may exhibit keen eyesight and hearing and, in some cases, the ability to detect other environmental factors such as barometric pressure and magnetism. However, most carnivores, seals and hoofed mammals have a strong dependence on visual sensing, although, as always, they display considerable

Red Deer hind with new-born calf

Harvest Mouse with one-day-old young

variation – some nocturnal or subterranean mammals, for example, have little more than the ability to distinguish light from dark.

The arrangement of three inner ear bones helps define mammals as a group, and leads to hearing being a well-developed sense in many mammals. This helps them to detect both predators and prey, and, in some species, enables long-distance communication. Several mammal groups have evolved extremely sophisticated hearing apparatus for echolocation – a form of sonar, whereby a mammal makes a noise and receives information about its surroundings by the echoes reflected back from nearby solid objects. It can be used both for navigation and detecting food. Bats rely on echolocation for much of the time, but many cetaceans (such as beaked whales) also use it in the murky ocean depths, and some shrews are also known to echolocate as a guide to moving around their environment.

Teeth are another feature of the mammalian skull that separates them from their reptilian ancestors: mammals have evolved differentiated teeth, with different shapes and functions. Most mammals also have two sets of teeth in their lifetime, the 'milk teeth' of youth being replaced by the adult complement, although this is not universal. As a general rule there are four types of teeth – incisors, canines, premolars and molars, the numbers of each on the upper and lower jaw constituting the dental formula. This, and the shape and size of individual teeth, can be extremely useful in distinguishing closely related species of, for example, bats and rodents.

Differences in dentition reflect the variety of food consumed by mammals. For example, different bats eat different types of insects, and this is reflected in the structure of their teeth. On a broader scale, a carnivore's teeth are profoundly different from those of a herbivore. Dolphins have teeth, but some whales have lost theirs completely, relying instead on plates in the palate to act as a filter for small marine organisms. But diet also leads to other morphological adaptations. Most carnivores have a relatively short, simple alimentary canal, while a herbivore's gut is typically long and convoluted, with separate chambers, in which the hard-to-digest cellulose of plant tissues can be utilised during a long passage (sometimes repeated) through the gut, usually aided by the actions of a community of intestinal microbes.

11

The history of Britain's and Ireland's mammals

In prehistoric times, in what is now Britain and Ireland, mammal-watching would have been a very different prospect from that of today. A cast-list of impressive megafauna has come and gone within the last 2 million years, as a succession of ice ages and interglacials alternately made the land suitable for Woolly Mammoths and Arctic Foxes on the one hand, and warmth-loving hyenas and macaque monkeys on the other. Many exciting mammals of old trod the ground now covered by the tide of humanity: in London the bones from a 125,000 year old Hippopotamus have been found beneath Trafalgar Square, Woolly Mammoth remains down the Strand, and Woolly Rhinoceros sub-fossils under Battersea Power Station. At various times Cave Lions (probably a subspecies of present day lions), Sabre-toothed Cats, Cave Bears, Giant Elks and Wild Horses roamed the landscape, and the remains of at least five species of elephant-like animals and four rhinoceroses have been dug from British soil.

We are mammals, of course, and among the prehistoric arrivals were relatives of ours. It is thought that the first humanoids arrived in Britain about 1 million years ago, in the form of *Homo antecessor*. Neanderthals *Homo neanderthalensis* were present up to about 40,000 years ago, but were replaced by our own species, *Homo sapiens*, of which there is evidence of colonisation from around 33,000 years ago.

Britain became isolated from the European mainland during the Mesolithic (Middle Stone Age) period, some 8,100 years ago, by post-glacial sea level rise, perhaps rather suddenly as a result of a tsunami triggered by landslides in Norway. Although becoming warmer, the climate had not yet become amenable for a number of species to reach as far north as the peninsula that became Britain; these include several widespread European species, such as Orkney (Common) Vole, Lesser White-toothed Shrew, Beech Marten and Garden Dormouse. These are all naturally absent from Britain, although the first two have subsequently been introduced.

Ireland, cut off from mainland Europe very much earlier (about 12,000 years ago), has an even more depleted fauna. Several common British mammals, such as Weasel, Mole, Common Shrew and Water Vole are absent from the island of Ireland, and several others that occur there now, including Bank Vole and Pygmy Shrew, were probably introduced, as has been the Greater White-toothed Shrew.

So, which animals were present at the very start of Britain and Ireland's new identity? Woolly Mammoths and rhinos had disappeared, with the last known records about 12,000 years ago, but some of the tundra animals held on longer. Reindeer and Wild Horse (Tarpan) occurred up to at least 8,400 and 9,400 years ago respectively, and could conceivably have occurred on higher ground much later. Younger Mesolithic sites show evidence of a still impressive suite of large mammals, including Red Deer and European Roe Deer, Wild Boar, Elk (Moose), Wolf, Brown Bear, Lynx and Eurasian Beaver. The smaller mammal fauna included Hedgehog, Weasel, Wood Mouse and Bank Vole, along with the cold-adapted Root Vole and the Pika, both of which subsequently died out here.

Also common from Mesolithic sites is the Aurochs, the ancestor of most domestic cattle, which inhabited forest and grassland throughout much of Europe. The last British record was about 3,000 years ago, but it was a familiar and important supply of meat for our

Brown Bear is believed to have become extinct in Britain in early Roman times.

*The **Lynx** was hunted to extinction in Britain, and last recorded in about 200 AD.*

ancestors and frequently features in cave paintings; it became globally extinct in 1627 when the last one died in Poland.

Many of the common post-glacial mammals of Britain and Ireland were woodland species, and the gradually warming oceanic climate drove the expansion of wildwood, which reached its greatest extent about 6,000 years ago, covering southern Britain and blanketing even parts of the Highlands. Animals such as Red Squirrel, Pine Marten and Wildcat would have been widespread, and the pre-agricultural Mesolithic people would have hunted forest animals. Forest-dwelling bats, such as Bechstein's Bat and Noctule, probably reached their highest populations during this period.

But as Neolithic cultures spread, their agricultural way of life necessitated the start of forest clearance, initially using stone tools. As natural woodland was gradually replaced by open semi-natural habitats, including grassland, heathland and downland, so the mammalian fauna changed, to the benefit of open ground mammals such as Field Vole.

Another important change commenced in Neolithic times: the widespread introduction of domestic mammals. Dogs were already here, used for hunting during the Mesolithic period. Woodland clearance and the introduction of farming, however, saw the appearance of a new suite of domestic animals that quickly became part of the landscape: goats, sheep, cattle and pigs. Goats and sheep originated in the Middle East, and arrived here around 5,400 years ago. Cattle had already been domesticated from the Aurochs and could no longer interbreed with their much larger ancestor, and there is clear evidence of their existence in Neolithic sites. Pigs were also present, their archaeological remains distinguished by being slightly smaller than Wild Boar.

The last of our familiar domestic hoofed mammals to be introduced was the horse. Its wild ancestor, the Tarpan, which was present in Britain until about 9,000 years ago, was domesticated on the plains of central Asia and eventually introduced here; the earliest confirmed record is from about 3,800 years ago, during the Bronze Age.

Whilst these domestic animals were brought to Britain intentionally, a number of other mammals also arrived as an unintended consequence of human movements and trade. The House Mouse, like sheep and goats, probably originated from the Middle East where it had become largely commensal with agricultural communities. It may have arrived here with Neolithic farmers, but the earliest records are from the Iron Age, approximately 2,800 years ago, around the same time as the domestic cat, brought in as an early biological control agent. Ship Rat appeared in Roman times, around 2,000 years ago and Common Rat much later, probably in the 18th century.

With the arrival of large numbers of domestic grazing animals, the larger predators in the countryside inevitably began to be seen as a threat to livestock, which triggered a long history of persecution. Brown Bears were the first to be eliminated, probably early on in Roman times (43 AD onwards), followed by the Lynx. The Wolf held on for longer, with the last one thought to have been killed in 1621. Other predators, such as Fox, Weasel, Stoat, Polecat and Badger would have also suffered, and still do to some extent.

By the time the Romans arrived, Britain bore little resemblance to the island first occupied by Mesolithic settlers. Much of the forest had been replaced by grazing land and arable fields, with numerous settlements. The new settlers continued to change the landscape, through road-building and the drainage of marshland. The Fens in East Anglia, no doubt with large populations of Water Voles, Water Shrews, Otters and Eurasian Beavers, were greatly reduced in extent. Woodland clearance continued apace through Saxon and Viking times, and indeed, when the Norman Domesday Survey was carried out in 1086, only 15% of England was under forest cover. Brown Hare, probably introduced during the Bronze Age, was one of the beneficiaries of the new landscape.

The Norman invasion brought several new mammals. At first, the Rabbit was kept confined in warrens to supply meat and fur, often on offshore islands (the first definite mention is from the Isles of Scilly in 1176) and later in parks. Warrens were carefully managed and guarded from predators; only much later (from about 1750) did the Rabbit become ubiquitous. With Rabbits came Ferrets (domestic Polecats), the latter to catch the former, the first records of ferreting dating from the 13th century.

The Normans also established at least 143 'forests', areas of land set aside for hunting of deer and Wild Boar by noblemen (90, including the New Forest in Hampshire, were reserved for royal use), and numerous parks, created by surrounding country estates with ditches and banks and stocking them with free-roaming, semi-wild animals, largely to provide a supply of meat. Parks and forests suited Fallow Deer, brought from southern Europe, which thrived and quickly formed established populations (or, arguably, 're-established', given the presence of identifiable Fallow Deer remains in interglacial deposits).

The Norman emphasis on hunting and husbandry led to legal and cultural changes regarding Britain's mammals. Game legislation introduced in the Norman period persists in modified form today. Game became property and, at the same time, the control of 'vermin' became institutionalised. This began to have a serious effect on many of Britain's carnivores. The Wildcat, for example, disappeared from southern England in the 16th century and from Wales by 1880. Both Polecat and Pine Marten also declined, and continued to do so until the late 20th century.

Persecution was not confined to the land. There is no doubt that seals were hunted in many parts of Britain, and were once certainly much more widespread than they are today (for instance, there were seal colonies on the Isle of Wight). Indeed, the first protection law for any British mammal was the Grey Seals Protection Act of 1914. The main marine casualty, however, was the North Atlantic Right Whale. Once widespread and numerous, it was hunted from Norman times, and especially between 13th and 17th centuries. The toll was so great that by 1700 it was hardly ever caught, and despite complete protection from the 1950s it has never recovered. Although there have been occasional sightings since then, this whale remains effectively extinct in the north-east Atlantic. It joins the Grey Whale (by 500 AD) and possibly the Walrus (Bronze Age) as sea mammals extinct in, or now very infrequent vagrants to, British or Irish waters.

Although persecution and exploitation of mammals were the dominant human influences from the 13th to 19th centuries, other factors that rendered the landscape increasingly unsuitable started to come into play. The numerous Enclosure Acts, especially from the start of the 18th century, led to a loss of common rights and open field systems. While this left a valuable legacy of hedgerows, it enabled farming – and the persecution of 'vermin' – to become ever more efficient. The increasingly industrialised landscape and higher human population continued to push many mammals to the fringes. Even deer, once so prized by nobility, became very rare. Red Deer and European Roe Deer became extinct over most of England by the end of the 18th century, and in Ireland Red Deer disappeared from all but the extreme south.

The Victorian era saw an increase in interest in natural history, including mammals, but the fashion for keeping a menagerie led to the introduction of several exotic species. Sika was introduced in 1840, Reeves' Muntjac in 1901, and, arguably most damagingly, the Grey Squirrel was released repeatedly between 1876 and 1929, with serious consequences for the native Red Squirrel. Later introductions and escapes included the Edible Dormouse (1902), American Mink (1929), Coypu (1929) and Muskrat (by 1930). While the aquatic Coypu and Muskrat have now been eradicated, the American Mink continues to wreak havoc, especially on the Water Vole.

The late 20th and early 21st centuries have, by comparison, been better times for Britain's mammals: an era of nature conservation and tree planting, and a movement towards appreciation rather than exploitation. The Wildlife and Countryside Act 1981 and subsequent domestic and European legislation have given mammals more protection than ever before, while the membership of wildlife conservation groups has never been higher. Otter, Polecat and Pine Marten have all shown significant range increases. With a trend towards 're-wilding', Eurasian Beaver has been reintroduced, and there are tentative plans to try and re-establish Lynx and even the Wolf. While many threats remain, not least anthropogenic climate change, the mammal fauna of Britain at least has a glimmer of hope for the future.

*The last British **Wolf** is believed to have been killed in 1621.*

Mammals – explaining the names and classification

All living species – millions of them – have been given unique names, so that we can recognise them, write about them, categorise them and describe their relationships: the names themselves are structured so as to describe how one species may be related to others in evolutionary terms. These 'scientific' names, strictly controlled by internationally agreed rules, have two words, based on Latin or Greek. Species are grouped into a larger category of related forms, a 'genus'; the first word is the genus (*e.g. Sorex*, the typical shrews). Adding the second word gives a unique combination for every species. A subspecies, or race, necessitates a third word. These names are always in *italics*, and have a capital initial for the first word only, never for the second or third.

The classification of living things is continually changing in light of new findings by scientists. Although various classifications have been proposed for mammals, the most widely accepted order of species is that proposed in *Mammal Species of the World: A Taxonomic and Geographic Reference* (3rd edition) (2005), edited by D. E. Wilson & D. M. Reeder, published by Johns Hopkins University Press. This classification is followed in this book[1], and is summarised in the chart *opposite* for those families recorded in Britain and Ireland. (This includes two families that were present during the current geological period (the Holocene) – which began about 11,500 years ago – but for which the species concerned have since become extinct.) The table also shows the number of species recorded in Britain and Ireland within each family (including those represented only by 'ephemeral' introduced species, where these have bred).

Although mammal classification and taxonomy provides a useful means of presenting perceived evolutionary relationships, it does not always result in species that appear similar being listed close together. Since this book is essentially about identification, the approach taken is to arrange the species accounts so that direct comparison between similar species is possible, rather than in taxonomic order. However, a list of the species in strict taxonomic order is included at the back of the book (*pages 314–316*).

Except for more obscure groups, scientific names are not often used in everyday conversation, although botanists (and gardeners) and some entomologists often find it easier to refer to plants or invertebrates in this way to avoid confusion. With mammals, people usually say they have seen a Common Shrew, not a *Sorex araneus* for example – but English names lack precision. There is, therefore, a trend towards adding extra words to popular English names to improve this precision in a global context – hence Red Fox, European Rabbit or Western Hedgehog – but in a British and Irish context such additions are clumsy without adding clarity. Although a few such names are becoming increasingly familiar, when only one species is found and there is no risk of confusion, most people are content to refer to the species as simply as the Fox, Rabbit or Hedgehog. This approach is adopted in the key reference to British and Irish mammals, *Mammals of the British Isles: Handbook* – 4th Edition (2008) by S. Harris and D. W. Yalden, published by the Mammal Society. These names, and, for species recorded since 2008, the English names used by the Mammal Society (mammal.org.uk), are used in this book.

[1] The only departure from this treatment relates to the Physeteridae and Kogiidae (the sperm whales), where the Society for Marine Mammalogy (2014) is followed (marinemammalscience.org/species-information/list-of-marine-mammal-species-subspecies/).

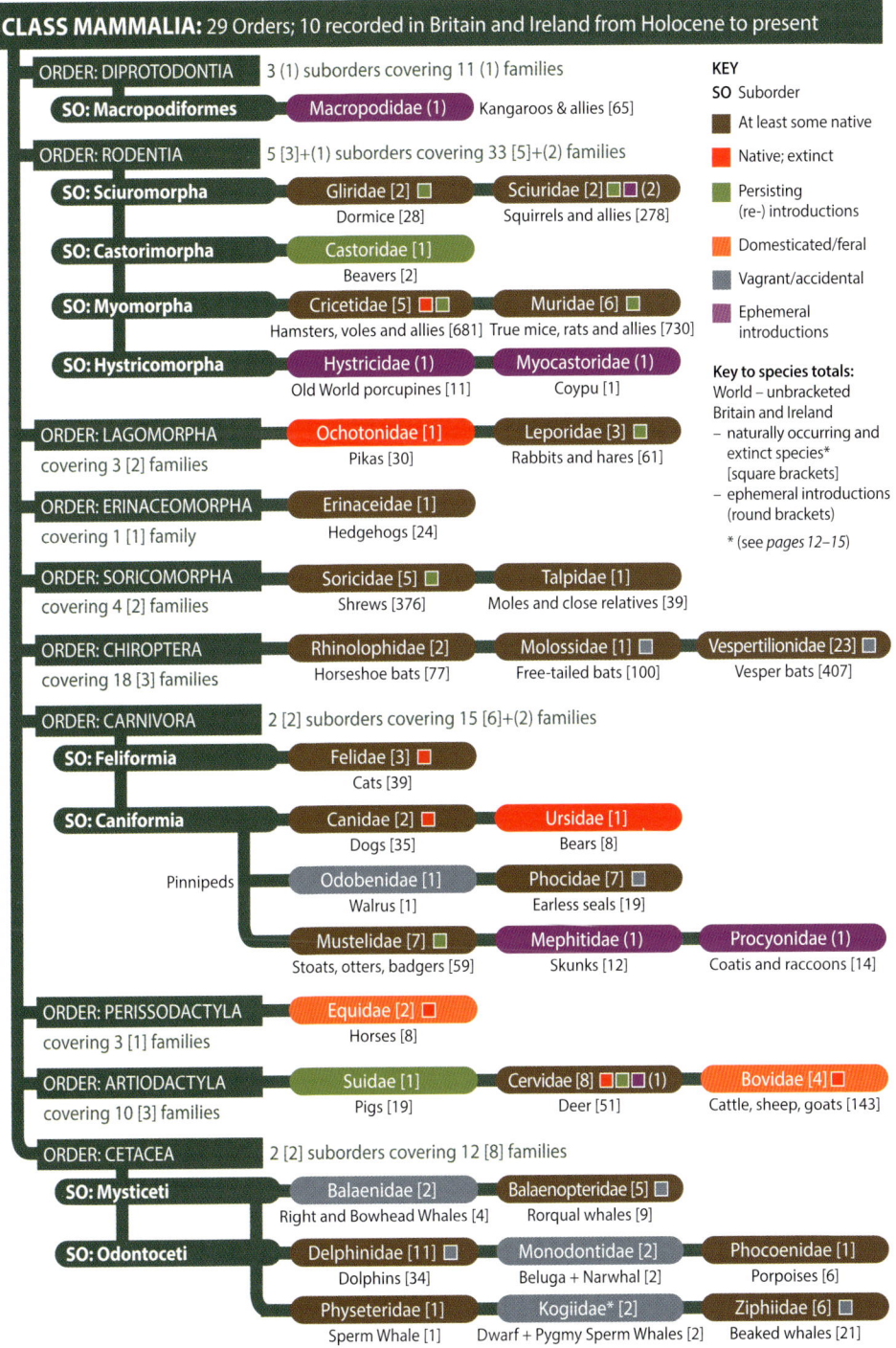

CLASS MAMMALIA: 29 Orders; 10 recorded in Britain and Ireland from Holocene to present

ORDER: DIPROTODONTIA — 3 (1) suborders covering 11 (1) families
SO: Macropodiformes — Macropodidae (1) — Kangaroos & allies [65]

ORDER: RODENTIA — 5 [3]+(1) suborders covering 33 [5]+(2) families
SO: Sciuromorpha — Gliridae [2] — Sciuridae [2] (2)
Dormice [28] — Squirrels and allies [278]
SO: Castorimorpha — Castoridae [1]
Beavers [2]
SO: Myomorpha — Cricetidae [5] — Muridae [6]
Hamsters, voles and allies [681] — True mice, rats and allies [730]
SO: Hystricomorpha — Hystricidae (1) — Myocastoridae (1)
Old World porcupines [11] — Coypu [1]

ORDER: LAGOMORPHA — Ochotonidae [1] — Leporidae [3]
covering 3 [2] families — Pikas [30] — Rabbits and hares [61]

ORDER: ERINACEOMORPHA — Erinaceidae [1]
covering 1 [1] family — Hedgehogs [24]

ORDER: SORICOMORPHA — Soricidae [5] — Talpidae [1]
covering 4 [2] families — Shrews [376] — Moles and close relatives [39]

ORDER: CHIROPTERA — Rhinolophidae [2] — Molossidae [1] — Vespertilionidae [23]
covering 18 [3] families — Horseshoe bats [77] — Free-tailed bats [100] — Vesper bats [407]

ORDER: CARNIVORA — 2 [2] suborders covering 15 [6]+(2) families
SO: Feliformia — Felidae [3]
Cats [39]
SO: Caniformia — Canidae [2] — Ursidae [1]
Dogs [35] — Bears [8]
Pinnipeds — Odobenidae [1] — Phocidae [7]
Walrus [1] — Earless seals [19]
Mustelidae [7] — Mephitidae (1) — Procyonidae (1)
Stoats, otters, badgers [59] — Skunks [12] — Coatis and raccoons [14]

ORDER: PERISSODACTYLA — Equidae [2]
covering 3 [1] families — Horses [8]

ORDER: ARTIODACTYLA — Suidae [1] — Cervidae [8] (1) — Bovidae [4]
covering 10 [3] families — Pigs [19] — Deer [51] — Cattle, sheep, goats [143]

ORDER: CETACEA — 2 [2] suborders covering 12 [8] families
SO: Mysticeti — Balaenidae [2] — Balaenopteridae [5]
Right and Bowhead Whales [4] — Rorqual whales [9]
SO: Odontoceti — Delphinidae [11] — Monodontidae [2] — Phocoenidae [1]
Dolphins [34] — Beluga + Narwhal [2] — Porpoises [6]
Physeteridae [1] — Kogiidae* [2] — Ziphiidae [6]
Sperm Whale [1] — Dwarf + Pygmy Sperm Whales [2] — Beaked whales [21]

KEY
SO Suborder
At least some native
Native; extinct
Persisting (re-) introductions
Domesticated/feral
Vagrant/accidental
Ephemeral introductions

Key to species totals:
World – unbracketed
Britain and Ireland
– naturally occurring and extinct species* [square brackets]
– ephemeral introductions (round brackets)
* (see pages 12–15)

Britain's and Ireland's mammals – an overview

Worldwide, there are about 5,000 species of mammal. In comparison, the mammal fauna of Britain and Ireland is rather impoverished, with just 24 native 'terrestrial' species, 17 species of breeding bat and two breeding seal species. Nine other bat and five seal species have occurred as rare migrants or vagrants. There are also 18 introduced or reintroduced land mammals and three semi-domesticated species. In addition, there are 30 species of whale and dolphin on the British and Irish lists, some of which are common, although many are very rare indeed.

As can be seen from the chart on *page 17*, the 122 mammal species that have been recorded in an apparently 'wild' state in Britain or Ireland during the last 50 years are representatives of 34 families. This section provides a summary for each of the 29 families that contain native species, cross-referenced to the individual species accounts.

The five mammal families that are represented only by non-native species are **Macropodidae** (Kangaroos and allies | Red-necked Wallaby, *page 261*), **Hystricidae** (Old World porcupines | Himalayan Porcupine, *page 266*), **Myocastoridae** (Coypu, *page 264*), **Mephitidae** (Skunks | Striped Skunk, *page 266*) and **Procyonidae** (Coatis and raccoons | Raccoon, *page 267*). The species concerned are distinctive and only one, the Red-necked Wallaby, has an established (albeit small) breeding population.

Raccoon

Red-necked Wallaby

Coypu

The following terms are used throughout the book to reflect the origin of each species:

Native:	occurs naturally.
Introduced:	accidentally or deliberately introduced by people within the last 1,000 years, now with self-sustaining populations.
Introduced (prehistoric):	introduced by people more than 1,000 years ago, now with self-sustaining populations.
Semi-domesticated:	introduced by people, often thousands of years ago, with the current populations being managed (or having been managed recently).
Migrant:	an individual that moves from one place to another to breed, feed or find shelter (in the context of mammals, a term restricted to bats and cetaceans).
Vagrant:	an individual that is outside its usual range or migratory route (in the context of mammals, a term restricted to bats, seals and cetaceans).
Accidental import:	presumed to have been imported accidentally by air or sea, although some species (several bats) may possibly occur as genuine vagrants.

THE TYPES OF MAMMAL

ORDER: Rodentia (Rodents)

The largest order of mammals worldwide. Small to medium-sized mammals, with short legs, four-toed feet, and often long tails. Feed largely by browsing and gnawing vegetable matter; teeth reflect this feeding habit – both jaws have a pair of chisel-like incisors with enamel only on the front, which continue to grow through the animal's life.

FAMILY: Sciuridae (Squirrels)   *Pages 54–61*

2 species: 1 native; 1 introduced (plus 2 ephemeral introductions)

Flexible, elongated body, bushy tail and rather large eyes. The back limbs are longer than the front, and have five toes; the fore feet have four toes. Extremely acrobatic, able to run up and down tree trunks and leap from branch to branch.

Red Squirrel

FAMILY: Gliridae (Dormice)   *Pages 62–65*

2 species: 1 native; 1 introduced

Dormice hibernate for up to seven months of the year, relying on the large accumulations of fat they put on in the autumn, when they almost double their weight. When active, they climb well and live mainly above ground, in the tree canopy. All the toes have sharp claws for climbing, and they have long, furry tails.

Hazel Dormouse

FAMILY: Cricetidae (Voles)   *Pages 69–79*

5 species: 3 native; 1 introduced (prehistoric); 1 extinct

Small, plump rodents, characterized by a blunt muzzle, small eyes and ears, and a relatively short tail. Move with a scurrying run on short limbs. Populations often show dramatic peaks and troughs. They are key prey species for many birds of prey and other mammals (*e.g.* Weasel).

Bank Vole

FAMILY: Muridae (Rats and mice)   *Pages 80–97*

6 species: 3 native; 3 introduced

Small to medium-sized rodents with pointed snouts and prominent whiskers, and long, scaly, almost hairless, tails. All British species have strong legs and feet, and are capable of hopping. Able to produce multiple litters each year.

Wood Mouse

FAMILY: Castoridae (Beavers)   *Page 98*

1 species: formerly native, reintroduced

Britain's largest rodent, the semi-aquatic Eurasian Beaver has large incisor teeth, so powerful that they can be used to gnaw and fell trees, modifying their habitat by creating dams and lodges. The flat, scaly tail and webbed hind feet aid swimming.

Eurasian Beaver

ORDER: **Lagomorpha** (Rabbits and hares)

Brown Hare

Almost strictly herbivorous, lagomorphs have four incisors in the upper jaw, with enamel on both the front and back faces, which grow continuously through their lives. Often eat their own droppings to gain maximum nutrition from their plant food.

FAMILY: Leporidae (Rabbits and hares) *Pages 116–127*
3 species: 1 native; 2 introduced

Long ears and an extended muzzle, with large eyes on the side of the head for all-round vision. Long, powerful hind limbs enable them to run at considerable speed. The fore feet have five toes and the hind feet four.

ORDER: **Erinaceomorpha** (Hedgehogs)

Hedgehog

All European species have spines, sharp teeth for their omnivorous diet, and five toes on each foot. Sometimes grouped with the Soricomorpha as the order Eulipotyphla.

FAMILY: Erinaceidae (Hedgehogs) *Page 112*
1 species: native

Britain's only spiny mammal. The spines are modified hairs, an adaptation for defence against predators. Hedgehogs have a long head and snout, relatively large eyes, short legs and a rotund body, with almost no tail.

ORDER: **Soricomorpha** (Shrews and moles)

Small mammals with five toes on each foot, and sharp, spike-like teeth for catching animal prey (usually invertebrates).

Common Shrew

FAMILY: Soricidae (Shrews) *Pages 100–109*
5 species: 3 native; 2 introduced

Small mammals, sometimes confused with mice, but with a long, pointed snout, very small eyes and short legs. Very active and highly predatory, shrews have sharp teeth for catching and killing invertebrates. They live mostly on the ground, or in leaf-litter.

Mole

FAMILY: Talpidae (Moles) *Page 110*
1 species: native

Well adapted to life underground, with a cylindrical body, spade-like front limbs with sharp, powerful claws for digging, tiny eyes hidden in the velvety fur, and large, sensory whiskers. The long snout has rows of sharp teeth, for catching and eating earthworms and other invertebrates. Creates extensive tunnel systems, with obvious piles of excavated earth known as 'molehills'.

ORDER: **Chiroptera** (Bats)

The only mammals capable of powered flight, the front limbs of bats are modified into wings, with four extended digits supporting broad membranes of skin. The thumb is free, while the hind limbs project backwards and join up with the wing membrane. At rest, hang upside down by the feet, which have a special locking mechanism to maintain a firm grip. Most (all British species) are nocturnal and eat insects, using echolocation to find prey and navigate around their environment. Bats hibernate during the colder months of the year when flying prey is scarce. The second largest order of mammals after the rodents, representing about 20% of all mammal species worldwide.

FAMILY: Rhinolophidae   *Pages 140–145*
(Horseshoe bats)
2 species: native

Horseshoe bats are distinguished from other bats by their 'nose-leaf' (complex folds of skin on the face), from which their ultrasonic echolocation signals are produced. They also wrap their wings around the body when at roost.

Greater Horseshoe Bat

FAMILY: Vestertilionidae (Vesper bats)   *Pages 146–193*
23 species: 16 native; 7 vagrants (plus 3 accidental imports)

Most British bats are from this family, members of which lack a complex, fleshy 'nose-leaf', and fold the wings (rather than wrapping them around the body) when hanging at roost.

Brown Long-eared Bat

FAMILY: Molossidae (Free-tailed bats)   *Pages 190, 194*
1 species: vagrant (plus 1 accidental import)

A family of bats characterized by much of the tail projecting beyond the tail membrane.

European Free-tailed Bat

ORDER: **Carnivora** (Carnivores)

Showing a very wide range of size and form, carnivores have a number of features which reflect their wholly or partly carnivorous diet. They have strong, sharp claws, with four or five toes on each foot, and characteristic dentition: on each jaw, six incisors, two well-developed conical canine teeth, and two of the cheek teeth (carnassials) are blade-like, closing with a shearing action to shred meat.

Wildcat with Feral Cat hybrid features

Fox

FAMILY: Felidae (Cats)   *Pages 198–201*

3 species: 1 native; 1 feral; 1 extinct

Carnivores with rather flattened faces and domed heads; retractable claws and soft toe pads; large ears; and big eyes that confer excellent night vision.

FAMILY: Canidae (Dogs)   *Page 202*

2 species: 1 native; 1 extinct

Characterized by a long, bushy tail, long legs, and four-toed feet (a fifth toe is vestigial) which do not have retractable claws. They are omnivorous and often notably sociable.

Polecat

Common Seal

FAMILY: Mustelidae   *Pages 206–227*
(Mustelids: otters, weasels and badgers)

7 species: 6 native, 1 introduced

Mostly long-bodied, with short legs, rounded ears and thick fur. They have anal scent glands that produce a strong smell used in sexual and territorial signalling. Apart from the Badger, all are largely solitary.

FAMILY: Phocidae (Earless seals)

7 species: 2 native; 5 vagrants   *Pages 270–279*

Marine carnivores, adapted for a largely aquatic lifestyle, with streamlined body and modified limbs (flippers) for underwater propulsion. A thick layer of blubber beneath the skin provides insulation and buoyancy, and acts as an energy store. Ungainly on land, using their belly and front flippers to 'caterpillar' along the ground.

FAMILY: Odobenidae (Walrus)   *Page 275*

1 species: vagrant

The Walrus is a massive seal-like marine mammal with large tusks. It feeds almost entirely on molluscs.

Walrus

ORDER: **Perissodactyla** (Odd-toed Ungulates)

Usually large herbivorous hoofed mammals, the hoof formed by an odd number of toes. In the case of horses, the hoof is formed from a single toe. They digest plant cellulose in their intestines rather than in one or more stomach chambers, unlike the even-toed ungulates.

FAMILY: **Equidae** (Horses) *Page 258*
2 species: 1 semi-domesticated; 1 extinct

Large and heavy-bodied, with long legs and neck, and long, flowing mane and tail.

Horse (Exmoor pony)

ORDER: **Artiodactyla** (Even-toed Ungulates)

Medium- to large-sized ungulates (hoofed mammals), the hoof formed of two central toes, giving a cloven-hoofed print. The other toes are small and raised, although may register as tracks on soft ground. Mostly herbivorous, with a multi-chambered fermenting stomach for rumination, allowing more efficient digestion of plant food.

FAMILY: **Suidae** (Pigs) *Page 226*
1 species: native, reintroduced

Pigs have a heavy body, relatively short legs and a very large head. The long snout ends with disc-shaped nose, and some teeth are enlarged into tusks. Unlike other Artiodactyla, pigs have a simple stomach, without ruminating chambers.

Wild Boar

FAMILY: **Cervidae** (Deer) *Pages 230–251*
8 species: 2 native, 4 introduced, 1 semi-domesticated reintroduction; 1 extinct

Deer are ruminant ungulates with four-chambered stomachs. They are generally tall with slender legs and long necks, and can run fast and swim well. Males (and females of some species) have antlers, the fastest-growing bones in the animal kingdom. They are shed once a year (unlike horns, which are permanent).

Red Deer

FAMILY: **Bovidae** *Pages 252–257*
(Cattle, sheep, goats)
4 species: 2 introduced (prehistoric); 1 semi-domesticated; 1 extinct

A large group of ruminant ungulates with, at least in the males, permanent horns made of keratin. They have four toes on each foot, but walk on the middle two, the cleaves.

Feral Goat

ORDER: **Cetacea** (Cetaceans)

MYSTICETI – BALEEN WHALES

Baleen whales have a series of baleen plates that hang, comb-like, from the upper jaw. They filter-feed by opening their cavernous jaws while swimming; when the mouth is closed water is strained through the baleen plates, leaving quantities of small fish and zooplankton trapped inside.

Bowhead Whale

FAMILY: **Balaenidae**   *Page 303*
(Right and Bowhead Whales)

2 species: both extremely rare vagrants from depleted North Atlantic populations

Right and Bowhead whales have a large head, up to a quarter of the body length, a broad, stocky body that lacks a dorsal fin, and a characteristic 'V'-shaped blow. They feed by 'grazing', swimming open-mouthed to create a continuous flow of water that passes into the front of the mouth; prey is trapped as the water is filtered through long, fringed baleen plates located along the mouth sides.

Minke Whale

FAMILY: **Balaenopteridae**
(Rorqual Whales)   *Pages 282–287, 300, 304*

5 species: 2 possibly resident; 1 regular visitor; 2 rare offshore

Rorquals have a streamlined body and large, flattened head. The blow is a single column. They are named after the rows of pleated and grooved folds of skin running from the throat to the navel, and feed by 'gulping' in water, which the throat pleats expand to accommodate. They then close their mouth and contract the pleats, which forces the water out through the baleen plates, trapping any prey inside.

ODONTOCETI – TOOTHED WHALES

This group includes all other cetaceans, from the huge Sperm Whale to the diminutive Harbour Porpoise. Rather than filter-feeding, they use various methods to locate and pursue their prey, predominately fish and squid.

Sperm Whale

FAMILY: **Physeteridae**   *Page 301*
(Sperm Whale)

1 species: offshore

The Sperm Whale has a very large, square head (out of proportion to the rest of the bulky body) and a narrow, underslung lower jaw lined with teeth. They do not have a dorsal fin, only a small 'hump'. The single blowhole on the top front of the head is offset and produces a blow that is angled forward and to the left.

Dwarf Sperm Whale

FAMILY: **Kogidae**   *Page 308*
(Pygmy and Dwarf Sperm Whales)

2 species: rare vagrants

Dolphin-sized, compact sperm whales with a characteristic blunt, squarish head and narrow underslung jaw lined with teeth. The dorsal fin is small and falcate in shape. Both species are poorly known on account of their timid nature and elusive deep-water habits.

Cuvier's Beaked Whale

FAMILY: Ziphiidae *Pages 302, 305–307*
(Beaked Whales)
6 species: all rare vagrants, often found stranded

Medium-sized whales with a distinctive protruding jaw, or beak. The forehead is often bulbous and houses the 'melon' used in echolocation. They have a streamlined body, with a dorsal fin located two-thirds of the way along the back. Notoriously difficult to identify and not well known owing to their unobtrusive behaviour and preference for deep water well offshore. They are known to dive to great depths in search of squid and fish.

Beluga

FAMILY: Monodontidae *Page 310*
(Narwhal and Beluga)
2 species: both rare vagrants

The unusual Narwhal and Beluga make up this family of medium-sized whales. They both lack a dorsal fin and have a prominent bulbous 'melon'. They have reduced teeth: in the case of the Narwhal just two, the left one of which forms the tusk. Both are extremely rare visitors from the Arctic.

Short-beaked Common Dolphin

FAMILY: Delphinidae
(Dolphins) *Pages 288–297, 299, 305, 309*
11 species: 6 resident; 2 regular visitors;
3 rare vagrants

The dolphin family includes two distinct groups: **Dolphins** are generally small, beaked cetaceans with streamlined, often distinctively patterned bodies, and tall, falcate dorsal fins located midway along the back; and **'Blackfish'**, that are generally much larger, predominately black, animals that only have a small beak at most. The dorsal fin is relatively large and prominent. Both groups are generally highly social and demonstrative, and capable of great speed and acrobatic behaviour.

Harbour Porpoise

FAMILY: Phocoenidae *Page 298*
(Porpoises)
1 species: resident

The smallest of the cetaceans, porpoises have plump bodies and a small beak with flattened, spade-like teeth, different from the conical teeth of dolphins. They are timid, unobtrusive animals that inhabit inshore waters where they feed on fish.

Watching mammals

Compared with mass-participation hobbies, such as birdwatching or angling, there are relatively few dedicated mammal watchers. Part of the reason for this is the nature of mammals themselves – they are shyer and more secretive than birds, and often less colourful and more nocturnal. But another reason is surely that many people do not realise that they can go mammal-watching and actually see things. Mammals have a reputation for being difficult to observe, and that is certainly true if you simply use 'birdwatching methods' to see mammals. But with a different approach, a whole new world opens up.

Almost all the animals in this book, with the exception of the rare bats and sea mammals, can be encountered relatively easily, in the right place and using the right methods. For many iconic British species, such as Badger, Otter, Water Vole and Bottlenose Dolphin, it requires little more than a trip to a known locality and patience to get views of wild individuals. Other species can be seen with twilight and night-time expeditions, while many small mammals and bats require specialized, but readily available, activities and methods such as live trapping, and checking caves and bat-boxes. And if all else fails, the observation of field marks and signs will often provide definitive evidence of the presence of a species.

Grey Squirrel on a feeder

The Red Deer rut can be a thrilling spectacle.

Mammal-watching techniques

If you wish to see some mammals on a 'normal' country walk, there are a number of commonsense steps you should take. Firstly, wear clothes of neutral colours and fabrics that will not make rustling sounds while you are moving about. Secondly, remain as quiet as you can, speaking in a whisper at best, walking on grass rather than gravel where you can, and turning your mobile phone to silent. Thirdly, try

Water Voles can be seen by sitting quietly on a riverbank and waiting.

to refrain from making sudden movements and walk steadily. And finally, on the whole it is best to go out alone or with just one other person. People in groups see fewer mammals than those who are on their own, although the requirements of safety and security, especially on evening or night walks, must take precedence.

Make sure you take binoculars for scanning and keep alert as to where to look. For example, if you are on a track, repeatedly check ahead as far as you can see; hard-to-see mammals such as Weasels and Stoats, as well as deer, Foxes and voles often run across tracks. Watercourses are also thoroughfares, and it often pays simply to sit quietly by a riverbank in the hope that either an Otter or Water Vole will swim by, or that any mammal will simply come along a towpath. The edges of fields, especially near cover, are also good places to look for mammals, as are the sides of dry stone walls. If you do see a wild mammal, try to keep downwind: many mammals can easily detect humans by scent, causing them to flee or hide before we have chance to see them.

Spotlighting

Much mammal-watching is done at twilight and in the dark, so a particularly useful way to find mammals is by spotlight, using a powerful torch that can illuminate, for example, across a field. Spotlights can cause disturbance to mammals and other wildlife, and should therefore be used sparingly; fitting a red filter over the light will cause less disruption to the night vision of the animal. Spotlighting can be done on foot or in a slow-moving vehicle: simply move the beam back and forth as you move forward, looking for illuminated shapes and, more dramatically, reflected eyeshine.

Many mammals have a special organ, known as a tapetum, which reflects light back from behind the retina. This increases the light available to the photoreceptive cells, a useful adaptation to low light living, but which also produces visible eyeshine, often of a distinctive colour, although it will vary according to the angle from which it is observed. Both the colour and the estimated height above ground give some idea of the mammal being observed.

Rabbits show red eyeshine in a spotlight at night.

Badgers can sometimes be enticed into gardens at night to feed on peanuts.

Using a car

Vehicles can make excellent hides, and in some places (for example the plains of Africa) they are routinely used for watching wildlife, especially mammals. Going 'on safari' by car, even in Britain, can be productive. Set out at dusk and drive along country lanes slowly and considerately, looking ahead for animals crossing the road and around you. As night falls, the headlights illuminate mammals in the beam. This is a good technique for seeing many species, but when doing this it is important always to obey the Highway Code and ensure that the driver concentrates fully on the driving.

Lures and baits, and simply waiting

Some of the best mammal-watching can be done in the comfort of your own home. Gardens can be much better for mammals than might be expected, often hosting Foxes, Badgers, Hedgehogs, Wood Mice, Common Shrews, Common Rats and a range of bats. If you put out their favourite foods, many can be attracted, especially at night; ramps can be provided for small mammals to reach scraps on a bird table. Under low lighting, the mammals on your doorstep may reveal themselves.

All over the country, hides have been provided in nature reserves or on private land, from which it is possible to watch everything from Bank Voles to Pine Martens. Some hides overlook wild country where animals can come and go; others overlook sites where bait is left out, encouraging mammals to visit. Either way, the mammals are wild and the encounters can be memorable, but as always patience and perseverance may be needed.

Another technique for seeing mammals is to use a so-called 'high seat', a platform reached by a ladder. Originally designed for deer-shooting, such seats, or a natural approximation thereof like a cleft in a tree, can be useful for mammal-watchers. The observer is raised above the ground, where mammals do not expect a human to be, and where your scent may be wafted over their heads.

Camera traps

The use of camera traps (also known as trail cameras) to detect and study animals has increased as suitable devices have become more affordable. These cameras are activated remotely by the motion of an animal triggering a sensor, causing little disturbance to the subject. If used at night they can provide a fascinating insight into the habits of our little-known nocturnal wildlife.

When to go mammal-watching – time of day

All mammals show peaks of activity at some time during the day; unfortunately, many British species are more active during the night than the daytime. Even species that we often see when it is light, such as Brown Hares and Weasels, may fall into this category. The Brown Hare that you see sedately browsing at midday could be involved in excited mating chases at midnight. Any wildlife enthusiast who becomes a keen mammal watcher will inevitably make the shift towards being outdoors at night.

It is often at twilight when mammals and their watchers come face to face. Dusk twilight is the time when a nocturnal mammal may be tempted out early by hunger and a diurnal animal might be searching for its last meal before sleeping, with the pattern reversing at dawn. Some mammals have natural bursts of activity at twilight greater than at any other time of day, and are termed 'crepuscular'. Anyone who selects a quiet spot in twilight can expect some kind of mammalian encounter, even if only with a Rabbit, Fox or bat.

Getting to know mammals' activity peaks is an essential part of fieldcraft. Bat enthusiasts, for example, can predict with some accuracy when different species will appear in the summer night sky: Noctule when it is still quite light, pipistrelles not long after, Daubenton's Bat not until it is properly dark (see also *page 36*). But it is often more complicated than that. For example, Otters often feed at first light, and then take cover until they are hungry again, usually mid-morning. Cetaceans also often seem to be most active in the morning. The more you watch mammals the more you will come to understand their activity rhythms.

A Red Deer stag proclaiming his dominance at dawn.

When to go mammal-watching – time of year

Mammal watching can be rewarding at any time of year. You can enjoy the antics of squirrels in the middle of winter and watch bats flitting over the garden on a hot evening in June. Most mammals have annual rhythms that determine how active they are during each season, and these differ between species. Some British mammals hibernate, although fewer than most people imagine – really just bats, dormice and Hedgehogs. In fact, bats are readily roused from their hibernation and will hunt flying insects on mild nights, even in the middle of winter, although they soon return to torpor. Animals such as Badgers and Water Voles may become much less active and confined to their burrow in cold winter weather, but they do not enter long periods of torpor with a profoundly reduced metabolic rate, unlike true hibernators.

Winter

While winter is not without its charms, fewer mammals are active than at other times of year. Aside from the hibernators, some species are almost impossible to find. Shrews, for instance, do not hibernate, but their high metabolic rate means that mortality is high when it is cold. Almost all adults die over the winter, leaving their offspring to sustain the population the following year. Some small rodents, too, suffer from high mortality in cold winters: Harvest Mouse populations are usually very small by early spring. But more hardy rodents, such as Wood Mice and Field Voles, are active and often visit bird tables on winter nights.

Some larger animals are particularly active during this season. One of the sounds of mid-winter is the calling of Foxes, the dog's plaintive bark mixing with the vixen's squeal, best heard on a still, frosty night. December and January is their peak breeding season, the vixen being fertile for only about three days a year. December also sees a peak in the mating chases of Grey Squirrels and the main rut of both Chinese Water Deer and Wild Boar. The breeding season of Brown Hare often begins in February, and Badger cubs

Mountain Hares in their white winter coat can be conspicuous when there is no snow.

are also born this month, deep in the sett. Otters and Reeves' Muntjacs breed in any month, including mid-winter.

One bonus of the winter months may be snow. Suddenly we can see where mammals have been: fresh snow provides a good substrate for their tracks. It can be possible to follow tracks for a considerable distance, which is usually not possible in mud. Fresh snow with animal tracks is an exciting prospect, and you can treat it like a crime scene, trying to work out what went where and when. On the other hand, if there is a lack of snow in the hills, both Mountain Hares and Stoats may be unusually easy to spot, their white winter fur standing out incongruously against the dark background.

Spring

Most British mammals start their breeding season in March, April and May. In the case of rodents, shrews, Rabbits and Brown Hares, their first litter will be one of a succession, which gradually builds up the population through the summer. Brown Hares famously begin their 'Mad March' courtship which, contrary to popular belief, may occur right through the summer and autumn. This involves males chasing rivals and females alike, and the females 'boxing' unwanted attention away.

March is a good month to visit wetlands: the waterside vegetation has not yet grown up enough to hide animals such as Water Voles, Otters and American Mink. Likewise, deer can still be easy to see in woodland and on the edges of fields, without being obscured by a cloak of leaves. European Roe Deer antlers are shed late in the year and new ones have fully grown by March, while all the other deer shed their antlers in April and May and grow them through the summer until they are fully formed in August, ready for the rut.

As the spring takes hold, those animals that have hibernated begin to rouse: Hedgehogs by the end of March, bats by the end of April, and dormice in May. For some species,

The famous 'Mad March' courtship of Brown Hares is not confined to the spring, but is easier to see at that time of year as the vegetation has yet to grow tall.

ABOVE: *Hedgehogs emerge from hibernation in spring.*
BELOW: *Pods of spectacular Killer Whales (or Orcas) turn up each spring in the Shetland Islands.*

activity levels increase very quickly – Hedgehogs mate soon after they wake up, and bats form maternity colonies in May. Meanwhile, Badger cubs often make their first appearance above ground in late April, and May is the main month for deer to give birth to their young. When their time has come, female deer leave their social groups and find a quiet spot to give birth, the youngster being able to walk and run alongside its mother after just a few days.

Late spring can be a particularly good time to catch up with one of Britain and Ireland's most exciting mammals, the Killer Whale, mostly in Scottish waters.

Summer

Summer is the time of long days and long twilight, providing plentiful opportunities for mammal-watching. It is a good time to see such difficult species as Stoats and Weasels, which will have young and are more likely to be active. June is also the breeding season for Common Seals, although they are not colonial and the pups do not remain on land for very long. It is also the best time to watch Fox cubs playing outside the den.

The months of June, July and August are usually the peak period for cetaceans, a good time to take a ferry or special cruise across the Irish Sea, in the Hebrides or off southern Ireland. As well as dolphins and Harbour Porpoises, you could also find a Minke Whale, sometimes even from a headland viewpoint. The seawatching season lasts into September and October, although the later you are, the more likely it is that the sea state will be too rough for easy viewing.

June heralds the peak period for most bats to give birth. If the weather is poor at such a crucial time and adults are unable to catch sufficient insect prey, there can be significant infant mortality. In good weather, however, bats can seem to be everywhere, especially over water bodies and woodland clearings.

Young Foxes can be observed play-fighting near their den during the summer.

Typically, and out of keeping with the annual cycle of other deer, European Roe Deer begin rutting from mid-July and continue into August. Sometimes the males follow the females in close circles or 'figures-of-eight', producing distinctive worn 'roe rings' on the woodland floor. August is also the time when Hazel Dormice give birth to their young.

Autumn

Autumn is a bountiful time for encounters with mammals. Rodent populations reach their maximum, later than the midsummer peak of shrews, and it is the most productive time to set live-traps or attend a mammal-trapping session run by a local wildlife group (see *opposite*). Many mammals, especially the hibernators, are feeding actively to build up reserves for the winter, and the variety of cetaceans off our coasts is still impressive.

An autumn highlight, visually and audibly, is the deer rut. The rut of Red Deer, our largest land animal, is particularly dramatic, especially if you come across two stags locked in combat. Fallow Deer are also rutting, and have a bewildering variety of breeding systems, from single males searching for mates to large gatherings of bucks displaying within sight of one another. Sika may lack visual drama, but their extraordinary loud squealing calls are unlike anything else you can hear in the wild in Britain.

Another autumn spectacle is provided by Grey Seals. Unlike Common Seals, they form large colonies and produce white-furred pups at this time of year, which remain on land for several weeks. Some of the more accessible breeding grounds, such as Donna Nook in Lincolnshire, enable even those with limited mobility to have a memorable mammal experience.

More unusual delights are, as ever, offered by bats. Autumn is their main mating season, although the young are not born until the following summer. The males of some species hold territories and may even undertake 'song-flights', hoping to attract a female, or even a harem. Other species gather at 'swarming sites', usually at the entrance to a cave or tunnel, sometimes in substantial numbers.

Grey Seal pups can be seen on beaches from late autumn right through to December or January.

Small mammal trapping

A major difference between mammal-watching and birdwatching is that, sooner or later, the mammal enthusiast is likely to have to become 'hands-on', as you cannot realistically expect to see some species in any other way. Although only a small minority of birders take up ringing, many mammal-watchers quickly appreciate the benefits of using live-traps to study small mammals (mice, voles and shrews). However, it should be emphasised that the trapping of small mammals should never be undertaken lightly, and in the case of catching shrews it is illegal without a licence (see *pages 312–313*). As with all

Longworth Traps (RIGHT) and Tube Traps (LEFT) are used to catch and monitor small mammals for study purposes, and are usually placed on the ground.

animal interactions, the welfare of the mammal must be paramount: capture causes stress, and accidents do occur occasionally when live-trapping. But when carried out responsibly, live-trapping can deepen our understanding and appreciation of this difficult to study group, and provide valuable data on the distribution and biology of this neglected part of our fauna, a prerequisite for effective conservation action. Although the basic principles are outlined below, the Mammal Society has published a booklet, *Live Trapping Small Mammals* (see *page 317*), which provides much more detailed information. Nevertheless, for the sake of the animals' welfare and yours, it is important to learn how to use traps properly – either from someone with prior experience or by attending a formal training course, such as those run by, for example, the Mammal Society, local mammal groups or the Field Studies Council (field-studies-council.org).

There are several types of live-trap on the market, but they all consist of a baiting chamber and a tripping mechanism that closes the trap when the animal crosses it. The best-known traps are Longworth Traps, which have two separate parts: a tunnel, which contains the tripping mechanism, and a large chamber into which food, bedding (hay or cotton wool) and water is placed; this type of trap can therefore be left overnight without being attended. Some Longworth Traps have a shrew hole that allow these tiny animals to escape. It is important that whatever type of trap is used allows shrews to escape, unless they are being targeted specifically. It is essential to provide two kinds of food: nuts or seeds for vegetarians such as mice, plus some kind of meat (*e.g.* mealworms, castors (fly pupae) or cat food) in case you catch a shrew. If you anticipate that you might catch shrews you should ensure that the traps are inspected every two hours.

To catch a representative sample of the mammal fauna in the area that you are studying, the more traps that are deployed the better. Typically, 20 or 30 traps are placed along specific transects when undertaking a formal survey. Traps should be placed where they are most likely to be encountered by small mammals: for example alongside a log or branch on the ground, in a natural gap between grass stems, or at the base of a bush. As some species use thoroughfares above the ground, is also worth placing some traps along a tree branch.

Common Pipistrelle is the most abundant bat species in Britain and Ireland.

Bat watching

Identifying bats poses a particular challenge owing to their nocturnal and, to humans, apparently silent nature, the fact that they hide away during the day and the similarity of many species. Since all species are legally protected (see *pages 312–313*) and studying them closely requires a licence, you are likely to need specialist help if you want to get to grips with bats.

However, bat detectors provide an invaluable way of getting to know a little about these fascinating mammals. Many people appreciate bats for the first time when they see and hear a bat detector in use, perhaps on a bat walk organised by a local bat or mammal group or wildlife trust, and are surprised to find that there are different species of bat, often flying together. These detectors convert the echolocation ultrasounds made by bats into sounds audible to the human ear, and, because different species of bat use different echolocation frequencies, and the properties of their signals differ, flying bats can often be identified to species. You do not need a licence to use a bat detector or to enjoy listening to bats.

Nowadays there are many different types of bat detector available, although the most popular and least expensive are known as **Heterodyne Detectors**. They work in the way described above, 'translating' the sounds made by bats to our audible range. Most bats can be detected by tuning the dial to about 40 kHz; species identification then depends on the nature of the sound and the frequency at which it is loudest (see the summary on *pages 136–137*). To use this type of detector effectively requires practice but it is important to remember that not every species of bat can be conclusively identified using such a detector.

Other types of detector do not require tuning, as they scan the whole range of frequencies of the calls and produce a visible graph of frequency against time, which can be recorded for later analysis (see examples on *pages 134–135*). These include **Frequency Division**, **Time Expansion** and **Full Spectrum Detectors** and some can be left on for days at a time, recording the majority of calls that are detectable in a particular locality. Although such detectors are usually considerably more expensive than Heterodyne Detectors, by using them it is generally possible to distinguish between most species. In addition, apps are now available that display the calls in real time graphically on an iPad or similar device, and can identify the bat for you.

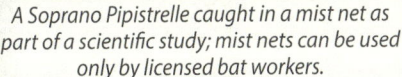

| A Soprano Pipistrelle caught in a mist net as part of a scientific study; mist nets can be used only by licensed bat workers. | Heterodyne Detectors are those most frequently used to find out which bat species are present. |

If you have an interest in bats, or want to find out more, it is worthwhile contacting your local bat group (see *page 319*). This might, for example, provide an opportunity to take part in counting bats as they **emerge** from roosts, or watch as **bat-boxes**, cellars or other sites are checked to see what is roosting there; sometimes, bats are caught and ringed and weighed, allowing close-up views. Licensed bat workers often visit hibernation sites in winter, such as tunnels or caves, in order to monitor overwintering populations and there may be an opportunity to assist in such surveys. Visits to cave or tunnel entrances are also made in the autumn, when **mist nets** and **harp traps** may be erected to catch foraging or **swarming** bats – swarming is when individual bats meet up, often having travelled some distance, possibly for the purpose of mating or introducing the young of the year to potential hibernation sites.

A Greater Horseshoe Bat being held, having been ringed and weighed prior to release.

Watching marine mammals

As with bats, but for very different reasons, the cetaceans are one of the truly 'difficult' groups of British and Irish mammals to identify. Living in the open seas, and spending much time below the surface, means they are usually seen only when you are specifically looking for them. Yet, with 30 species on the British and Irish list, cetaceans form our largest single group, and the diversity in our waters is among the best in Europe.

BRITAIN

1   Mull and the Inner Hebrides
2   **Ardnamurchan Point**
3   The Minches and Skye
4   Tiumpan Head
5   Orkney
6   **Shetland**
7   **Moray Firth**
8   Girdleness
9   Flamborough Head
10 Durlston Head
11 St Catherine's Breakwater
12 Portland Bill
13 Lyme Bay
14 The Lizard
15 **Isles of Scilly**
16 Gwennap Head
17 Cape Cornwall
18 Ramsey Island
19 Strumble Head
20 **Cardigan Bay**

IRELAND

21 Dún Laoghaire–Greystones
22 Hook Head
23 Ram Head
24 Co. Cork headlands
25 Loop Head and the Shannon Estuary
26 Achill Island

bold = prime sites

Key cetacean watching locations and ferry routes in Britain and Ireland

The easiest way to see cetaceans is to take a guided boat trip into suitable waters. Commercial trips are not particularly expensive, and the guides should be able to identify what you see. Being out in the animals' habitat also means that you often get close and sometimes spectacular encounters. But cetacean-watching is highly unpredictable, and to see a range of species, you may need to go out repeatedly. Only rarely are more than two or three species seen at a time, even in a hotspot such as western Scotland, Shetland or southern Ireland; and indeed, on some trips you may well see nothing at all.

It is also possible to see cetaceans by taking commercial ferries, a way of life in the Scottish islands, which can often bear fruit for the mammal spotter, especially in summer. Ferries have the advantage over smaller vessels in being more stable, and having high enough decks to allow you to scan over a longer distance.

Land-based observers do have a chance of encountering cetaceans too, but to maximize your chance of success you will need four things: 1) a panoramic viewing point over the sea, preferably on a cliff-top at the tip of a peninsula; 2) a telescope, or at least a pair of good binoculars for scanning; 3) a calm sea, so that any fins or bodies are conspicuous; and 4) patience. Even in the most productive waters, all you will see for much of the time is an expanse of water, birds and boats. However, the sight of a fin, a distant blow or even an animal breaching (see *page 281*), makes the wait worthwhile.

Whatever your viewpoint, a useful tip for spotting cetaceans is to look for flocks of seabirds. Where birds gather in feeding aggregations on the surface, fish-eating dolphins and porpoises may be in the water beneath, or the birds may be scavenging from the recent seal kill of a Killer Whale.

Seeing cetaceans is one thing, but identifying them, often just from a glimpse of their upper surface from a distance, or from a moving boat, is quite another. There is no quick fix to this; it is only by spending many hours in the field, preferably with an experienced observer, that you can translate sightings into identifications.

Tracks and signs

Mammals might sometimes be difficult to see, but they often leave signs of their presence behind, such as tracks, droppings and feeding leftovers. Such signs often prove invaluable as they can confirm the presence of an animal without the need to see it or trap it.

Since finding and identifying signs is such an important part of studying mammals, details of the most obvious field signs for each species have been have included in the relevant species accounts in this book. However, considerable experience is often needed to recognise mammal signs, and the subject is so broad that it warrants a book in itself (see *Further reading, page 317*).

The use of **tracks** in sand, mud or snow to identify and follow mammals has long been a part of hunting culture, used by people for millennia. Those same skills can be used to discover what mammals have been active in a locality. With a little practice, it is possible to recognise the distinctive two-toed prints of deer, the oval prints of a Fox or the almost human hand-like prints of Badgers. More experienced trackers may be able to tell the speed at which the animal was moving and, to some extent, what it was doing.

If you find a particularly well-formed track, it can be preserved by making a **plaster cast**. This is easy to do (details can be found in the books listed in the *Further reading* section or online) and you can even make your own library of prints. A different way of

Rabbit tracks in snow (hopping left to right)

Otter tracks in mud (walking left to right)

doing much the same thing is to put out **footprint tunnels**. These are triangular, tent-like tunnels into which bait is placed that can only be reached by the animal walking across an ink pad; as it walks out the animal leaves its inky footprints on special paper. This is a simple and non-invasive way to find out what is around; kits are inexpensive and can be purchased from, for example, the Mammal Society (see *page 319*).

Breeding sites can also be very distinctive. The holes of Badger tend to be arched over a flat base, while Rabbit holes are more oval and Water Vole holes are in groups just above the water surface (see *page 78*). Squirrel dreys in trees (see *pages 57* and *61*) and Harvest Mouse nests in reeds or grass (see *page 83*) are readily recognisable, as are the grassy nests on the surface made by Field Voles.

The entrance to a Badger sett has a characteristic shape, like a letter 'D' on its side, with the curve at the top.

Field Vole nests are situated close to the ground surface.

Mammals leave a wide variety of other signs. **Droppings** are often distinctive and, for many mammals, they are a form of territorial communication. It therefore pays for them to be recognisable, if only to their own kind. Fresh droppings may have a distinctive smell; those of carnivores are generally smellier than herbivore droppings. Placement of the droppings is important too. Some deer leave piles while others deposit as they go; Badgers dig special latrines; and Otters use various prominent places at the water's edge to mark their territory. For information on bat droppings, see *pages 138–139*.

Rabbit – round pellets (10 mm diameter). (Hare droppings are larger, more flattened and contain larger fibres.)

European Roe Deer – pointed pellets (10–14 mm long) – see opposite

Common Rat – pointed at one or both ends (12–18 mm long). (Mouse droppings are 3–6 mm.)

Hedgehog – cylindrical (10–50 mm long) and generally dark, with insect fragments.

Fox – very variable, but often long and pointed (50–150 mm × 20 mm), twisted at one end and containing fur, feathers, crushed bones and insect remains.

Otter – spraints (30–100 mm) have a distinctive smell and contain shell, fish scales and perhaps feathers and fur. They are often deposited in exposed positions as territorial markers.

Badger – droppings are very variable, but soft and often runny when the animal has been feeding on earthworms. Often deposited in shallow latrines, which are used to mark territorial boundaries.

Stoat – droppings (40–80 mm long) are twisted and contain fur, feather and bone fragments. Found on walls, stones and open ground. Weasel droppings are more twisted and smaller (30–60 mm long) on average.

IDENTIFICATION OF DEER DROPPINGS

Deer droppings are fairly easy to identify as such, being blunt at one end and pointed at the other. However, although size can be helpful, it is often very difficult to identify the species concerned as much depends on the size of the animal, and what they have been eating. Examples of droppings from four species are shown here for comparative purposes.

Reeves' Muntjac | 8–13 mm × 5–11 mm

European Roe Deer | 10–14 mm × 7–10 mm

Fallow Deer | 10–15 mm × 8–12 mm

Red Deer | 20–25 mm × 13–18 mm

Feeding signs include discarded nuts, browsed plants, and the remains of prey animals. The signs are almost as many and varied as the animals, and sufficiently distinctive often to be able to make definitive identifications. The corpse of a bird, for example, on which the flight feathers have been bitten cleanly through is the work of a Fox.

Small mammals leave clues to their presence by the way they open nuts such as hazelnuts.

Grey Squirrel: split into two halves

Hazel Dormouse: small, smooth-edged hole made in one side

Bank Vole: opened neatly, with tooth marks on inner rim

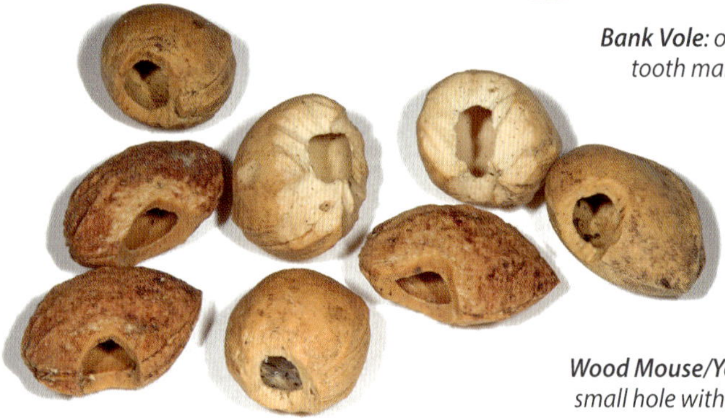

Wood Mouse/Yellow-necked Mouse: small hole with obvious tooth marks

Grey Squirrel (LEFT) *and Badger* (RIGHT) *often leave feeding signs by digging holes in lawns: squirrels in search of nuts buried the previous autumn, Badgers in search of leatherjackets (cranefly larvae).*

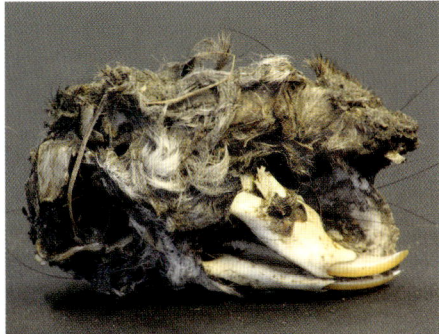

Badger hair caught on barbed wire *A Tawny Owl pellet with Common Rat remains*

Dead mammal bodies are signs in themselves. The presence of the more secretive species, such as Polecat, is sometimes revealed only by the remains of animals killed on the road. Mammal remains are often found in the regurgitated pellets of birds of prey and owls, and the study of such pellets provides invaluable evidence of the presence of particular species in an area (see *Further reading, page 318*). An excellent example of the value of such detective work came in 2007 when researchers in Ireland found the skulls of Greater White-toothed Shrew in Kestrel and Barn Owl pellets – the first evidence that this species was present on the island.

A browse line caused by feeding Red Deer.

Mammal tracks

Mammals often leave distinctive tracks in soft ground or snow, and illustrations of those that are most likely to be seen are presented here alongside each other for comparative purposes. Each print is from a full-grown animal and is shown roughly to scale within its group. For guidance, figures are given for either the average length or maximum/minimum range of a typical print for an individual species or group of species (group average or range shown in brackets). However, it is worth remembering that prints can look larger on softer substrates such as sand, or in snow as it begins to melt. A scale bar is presented with each group of prints, and a ruler is provided on the inside front flap of the book to enable you to measure the actual size of any print you find.

Rodents – four toes on fore feet (life-size)

Rats (pp. 92–97)

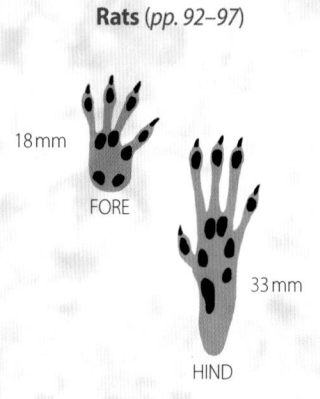

18mm
FORE

33mm
HIND

Common and Ship Rat prints are very similar, although those of adult Common Rat are larger. The long heel is distinctive, with six digital pads.

Squirrels (pp. 54–61)

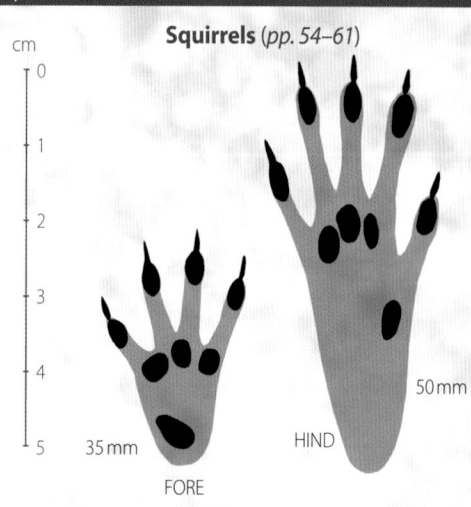

cm
0
1
2
3
4
5

35mm
FORE

50mm
HIND

Red and Grey Squirrel prints are indistinguishable. Smaller fore feet lie inside and slightly behind the hind feet.

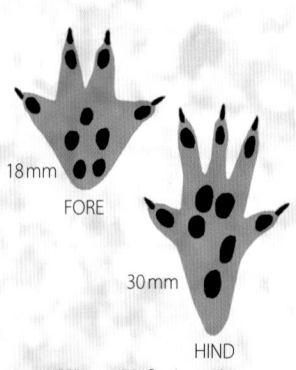

18mm
FORE

30mm
HIND

Water Vole (p. 76)
Star-shaped, splayed to 90°. Hind feet with five digital pads (six in rats).

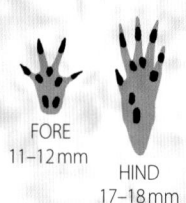

FORE
11–12mm

HIND
17–18mm

Other voles (pp. 69–75)
Field, Bank and Orkney Vole prints very similar. All broader than mouse prints.

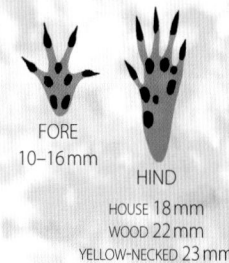

FORE
10–16mm

HIND

HOUSE 18mm
WOOD 22mm
YELLOW-NECKED 23mm

Mice (p. 80–91)
Wood, Yellow-necked and House Mouse prints similar to those of voles, but narrower.

Rabbit and hares (life-size)

Rabbit (*p. 116*)

Hares (*pp. 120–127*)

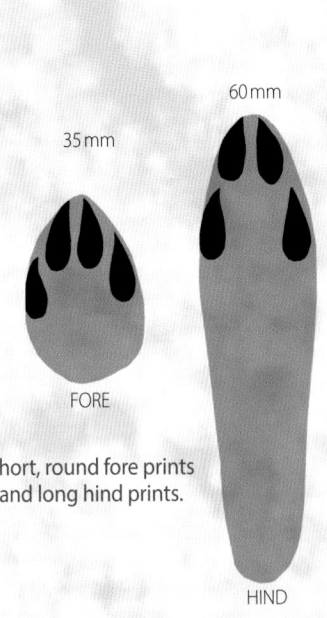

35 mm

60 mm

FORE

HIND

Short, round fore prints and long hind prints.

40 mm

FORE

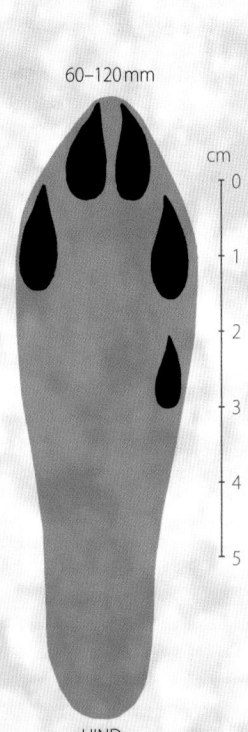

60–120 mm

HIND

cm

0

1

2

3

4

5

Relatively small, round fore prints and long, 'slipper-shaped' hind prints. Long hind feet often rest side by side behind the fore feet, leaving an impression like two upside down exclamation marks!

Shrews, Mole and Hedgehog (life-size)

Shrews (*pp. 100–109*)

Mole (*p. 110*)

Hedgehog (*p. 112*)

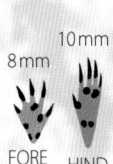

8 mm

10 mm

FORE

HIND

Small, faint; species usually indistinguishable. Five toes on fore feet (four in rodents).

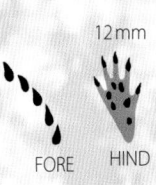

12 mm

FORE

HIND

Dotted lines made by claws on fore feet.

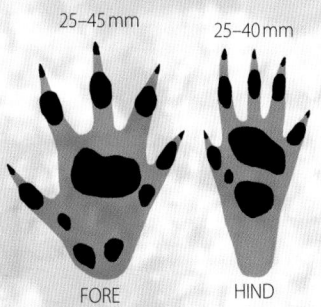

25–45 mm

25–40 mm

FORE

HIND

Fore feet more splayed than those of rats. Claws long; thumb often faint.

47

Carnivores (life-size)

Wildcat (*p. 198*)

40–60 mm

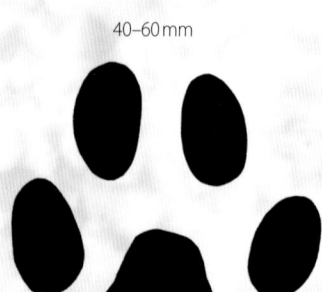

Fore and hind prints similar; larger than that of domestic cat. Claws do not show (retracted).

Feral Cat (*p. 201*)

35 mm

Almost circular; no claw marks.

cm
0
1
2
3
4
5

Fox (*p. 202*)

50–70 mm

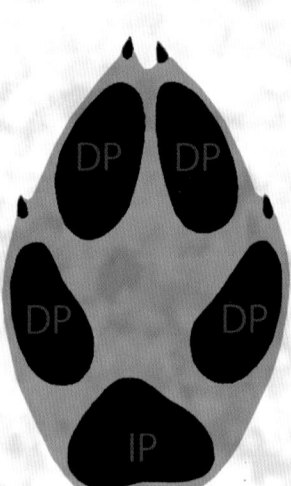

Small, oval. Two central toes well in front of the others. Interdigital pad (IP) no larger than the four digital pads (DP). Claw marks relatively long.

Dog

Very variable

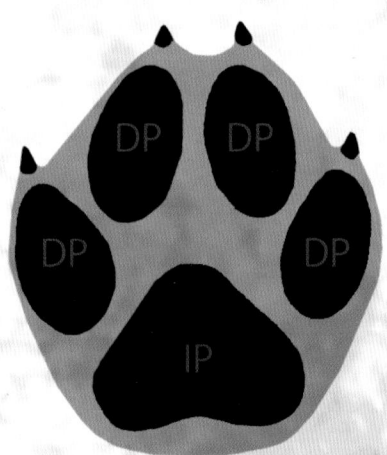

Like Fox but more rounded and with large triangular interdigital pad (IP). Size variable.

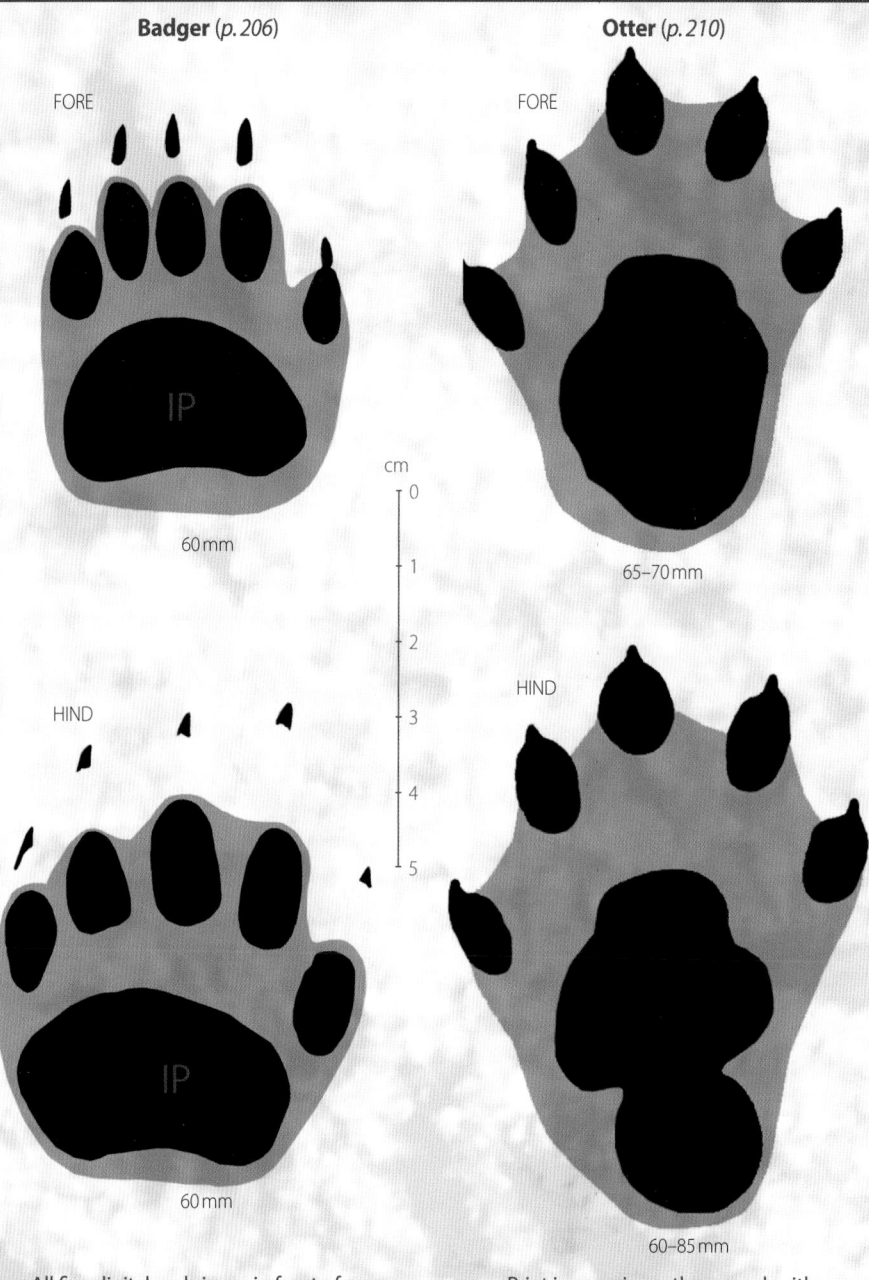

Badger (*p. 206*)

FORE

IP

60 mm

HIND

IP

60 mm

Otter (*p. 210*)

FORE

65–70 mm

HIND

60–85 mm

cm

0

1

2

3

4

5

All five digital pads in arc in front of very large interdigital pad (IP). Claws longer than similar carnivore tracks. Somewhat resembles a human hand.

Print impression rather round, with teardrop-shaped toes. Short claws point out from the digital pads. Webs present but are not always apparent.

49

Carnivores (life-size)

American Mink (*p. 214*)

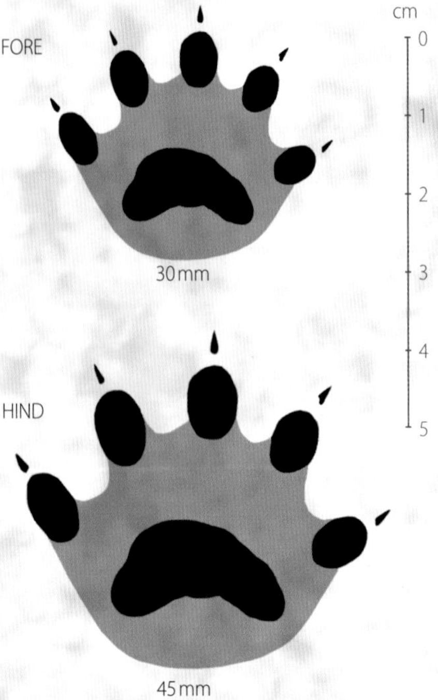

FORE

30 mm

HIND

45 mm

Fore and hind prints similar shape, often paired. Digital pads radiate out in star-like fashion; long claws.

cm
0
1
2
3
4
5

Pine Marten (*p. 216*)

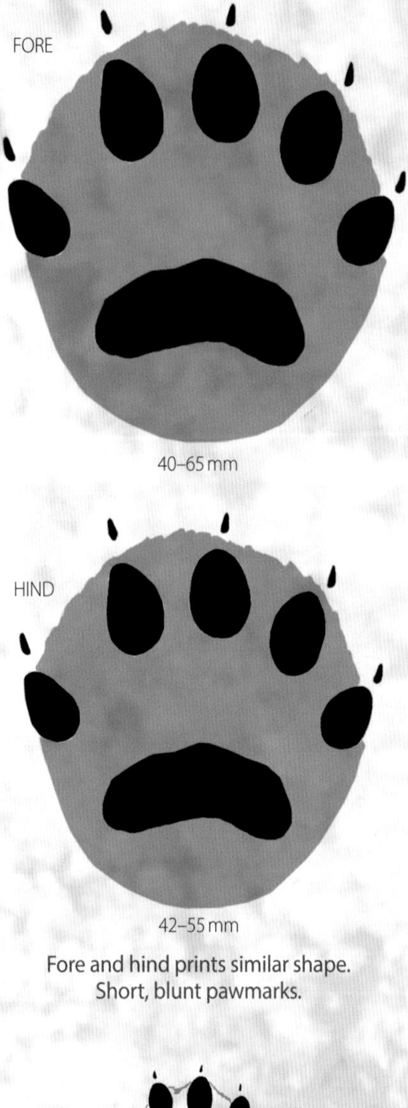

FORE

40–65 mm

HIND

42–55 mm

Fore and hind prints similar shape. Short, blunt pawmarks.

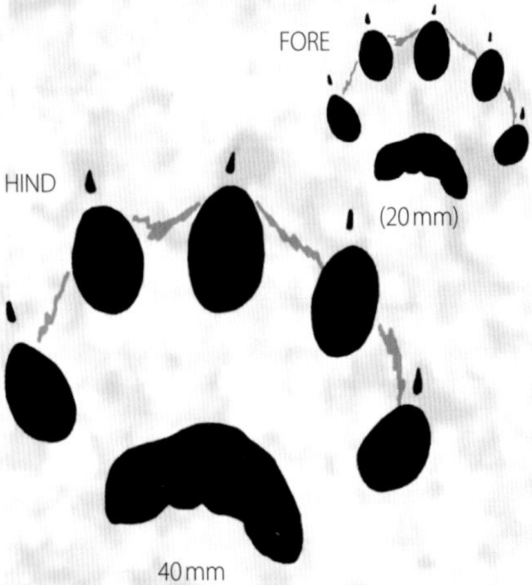

FORE

(20 mm)

HIND

40 mm

Polecat (*p. 218*) **and Stoat** (*p. 220*)

Fore and hind prints similar, hind approx. 2× fore. Small, sharp claws.

10–15 mm

Weasel (*p. 224*)

Shape as Polecat and Stoat, but smaller. Fore and hind prints similar in size.

50

Ungulates – Cow and Horse (half life-size)

Cattle (*p. 256*)

Horse (*p. 258*)

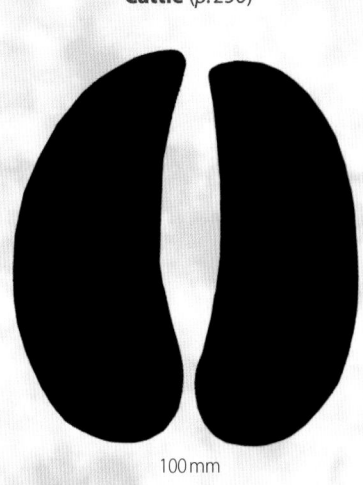

100 mm

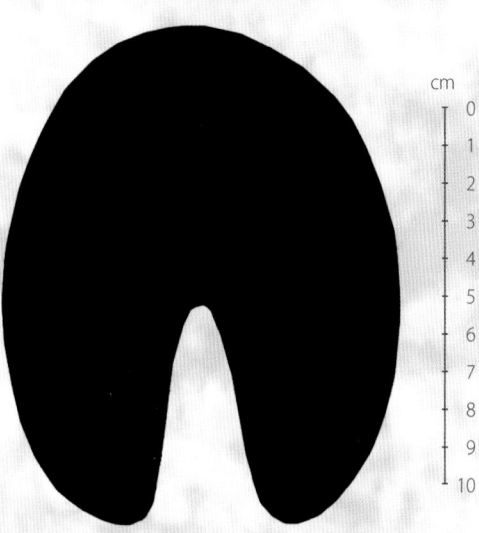

110–160 mm

cm
0
1
2
3
4
5
6
7
8
9
10

Cleaves are broad, slightly teardrop-shaped, and pointed at the front. Similar to those of sheep, but larger.

Free-living ponies unshod. Hoof size depends on size of animal. 'V'-shaped notch at rear.

Ungulates – Wild Boar, Feral Goat, Feral Sheep (life-size)

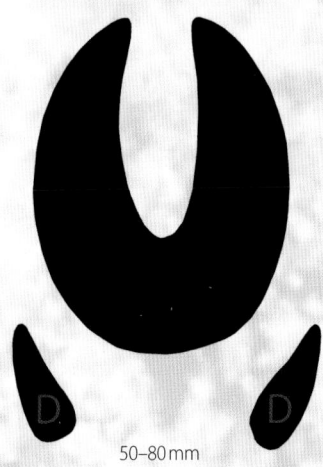

50–80 mm

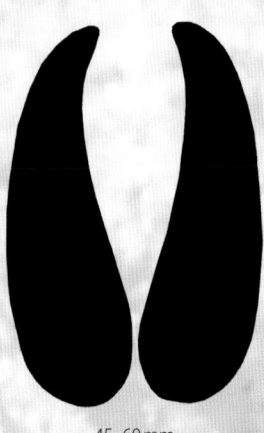

45–60 mm

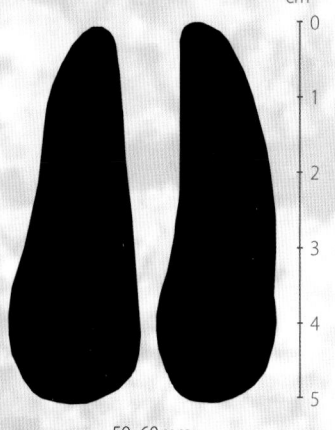

50–60 mm

cm
0
1
2
3
4
5

Wild Boar (*p. 226*)

Cleaves broad and rounded. Dewclaws (D) almost always register.

Feral Goat (*p. 252*)

Narrow and typically well-splayed at the front, rounded at the back.

Feral Sheep (*p. 254*)

Rather rectangular in shape, with broad cleaves rounded at both ends.

51

Deer (life-size) – print size very variable, depending on the age and sex of the individual

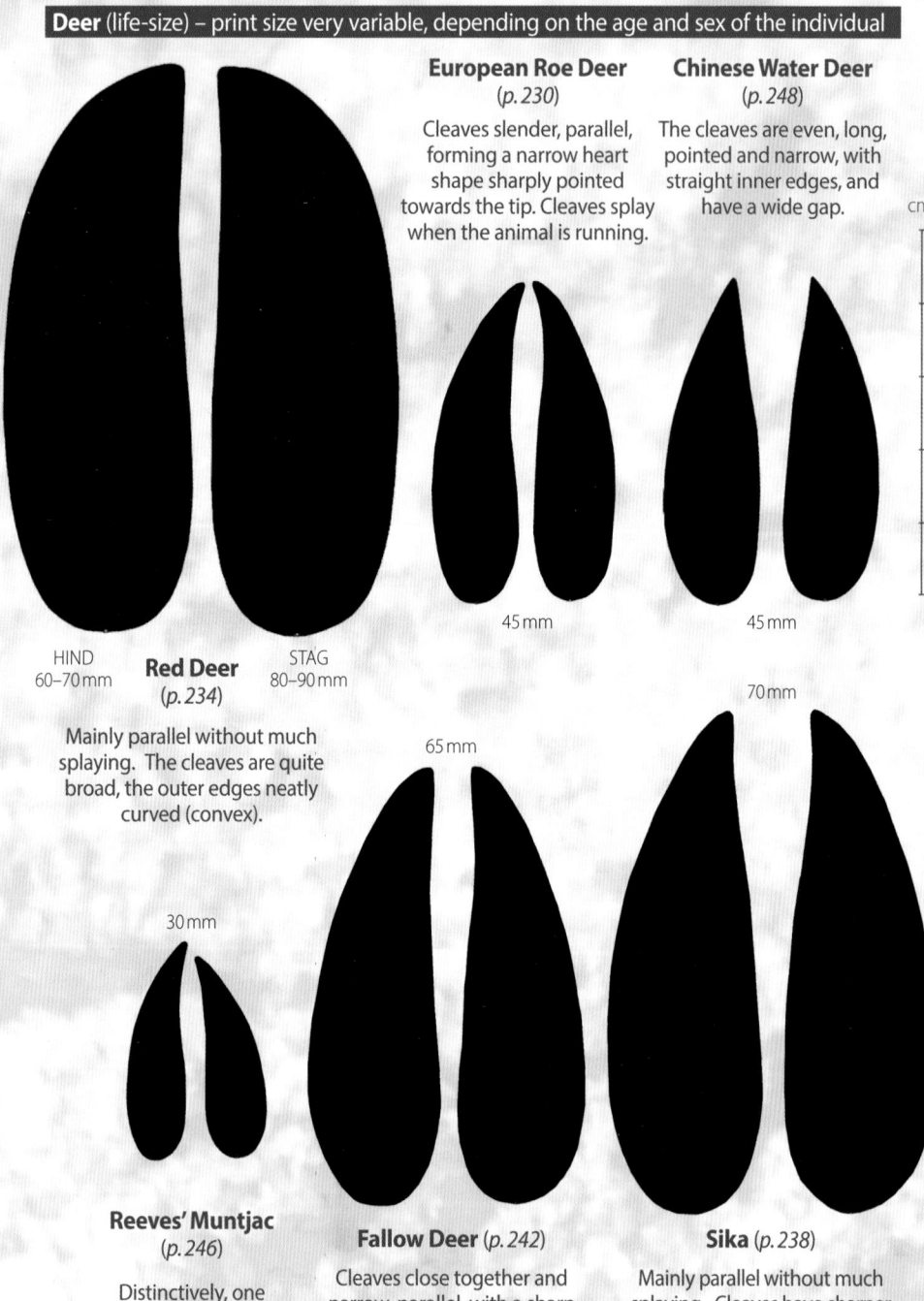

European Roe Deer
(*p. 230*)

Cleaves slender, parallel, forming a narrow heart shape sharply pointed towards the tip. Cleaves splay when the animal is running.

Chinese Water Deer
(*p. 248*)

The cleaves are even, long, pointed and narrow, with straight inner edges, and have a wide gap.

cm
0
1
2
3
4
5

45 mm

45 mm

HIND
60–70 mm

Red Deer
(*p. 234*)

STAG
80–90 mm

Mainly parallel without much splaying. The cleaves are quite broad, the outer edges neatly curved (convex).

30 mm

65 mm

70 mm

Reeves' Muntjac
(*p. 246*)

Distinctively, one cleave is always longer than the other.

Fallow Deer (*p. 242*)

Cleaves close together and narrow, parallel, with a sharp point; not splayed. More elongated than Red Deer tracks, narrower than Sika and larger than European Roe Deer.

Sika (*p. 238*)

Mainly parallel without much splaying. Cleaves have sharper points and are narrower and less curved than those of Red Deer.

The species accounts

Much of this guide is devoted to accounts for each species. These accounts provide a summary of key information on the mammal, with particular emphasis on identification, when, where and how to find it, biology and ecology, and, where available, population and status. Detailed information is provided for resident species and regular migrants, with rare migrants and vagrants (bats, seals and cetaceans) and introduced species being covered in a more concise manner. For some groups (small mammals, bats, deer and marine mammals) there is an introductory section to aid identification, and for difficult-to-identify pairs or groups of species, comparison spreads or summary tables. The order in which the species appear aims to enable direct comparison between similar species, rather than being strictly taxonomic (see *page 16*). The English names used in this book are those adopted in *Mammals of the British Isles* by Harris and Yalden (see *pages 16 & 317*).

NT **English Name**   *Scientific (incl. former) name* (see *page 16*)

Alternative name (where relevant)

◎ Pond Bat (*p. 193*)

LEGALLY PROTECTED

Biodiversity List (Sc)

Common*

HB:	4·3–5·5 cm
WS:	24–27 cm
Wt:	6–12 g

*Status boxes for marine mammals are dark blue

CONSERVATION STATUS IUCN designation (see *page 311*) is shown to the left of the English name. Codes used are: **EW** Extinct in the Wild **CR** Critically Endangered **EN** Endangered **VU** Vulnerable **NT** Near Threatened **DD** Data Deficient

◎ Highlights similar rare migrant or vagrant species (mainly bats and marine mammals).

LEGISLATION Red box if protected by domestic legislation; Orange box if included on the Biodiversity List for England (En), Wales (Wa), Scotland (Sc) or Northern Ireland (NI) (*pp 312–316*).

STATUS IN BRITAIN AND IRELAND An at-a-glance summary. Introduced mammals are indicated, but not so for native species.

MEASUREMENTS Typical adult biometrics: separate information is given for males (♂) and females (♀) where relevant. **HB:** head and body length **HS:** height at shoulder **WS:** wingspan **T:** tail length **Wt:** weight. For some species (bats & cetaceans), additional key information on morphology is presented in a separate box.

MAPS For Britain, maps for terrestrial mammals are based on data collated by the Mammal Society (see *page 319*) for 1995–2016 (■ shading) for the SNCB-funded Review of British Mammals project (see *page 319*), supplemented by the distributions shown in *Mammals of the British Isles* by Harris and Yalden (see *page 317*) for areas of low observer coverage (■ shading). Irish distributions are based on the *Atlas of Irish Mammals* (see *page 318*). Maps for marine mammals are from *Britain's Sea Mammals* by Dunn, Still and Harrop (see *page 317*), ■ indicating regular sightings and ■ occasional records. Records of vagrants or rare migrants are indicated by ●.

WHERE TO LOOK/OBSERVATION TIPS*

Suggestions of typical sites, key areas, best times and any other useful information that may help in obtaining a better encounter with the species.

NT Red Squirrel

Sciurus vulgaris

Once widespread, our native squirrel is now found only in northern England, Scotland, parts of Ireland and on a few islands (*e.g.* Isle of Wight and Brownsea Island, Dorset). The Grey Squirrel has replaced it in many areas, although the relationship is hotly debated: do Grey Squirrels cause the decline of Red Squirrels, or merely take advantage when the Red Squirrel population is hit by a setback such as the Squirrel Pox virus? Historically, the Red Squirrel was almost wiped out by deforestation in Scotland by the late 18th century, and was all but extinct in Ireland during the 17th century. It is likely that all Irish Red Squirrels, and many elsewhere, are descended from deliberate introductions.

LEGALLY PROTECTED	
Biodiversity List (En, Wa, Sc, NI)	
Locally common	
HB:	18–24 cm
T:	14–20 cm
Wt:	250–300 g

Distribution of established wild populations, for Britain based on records since 2010.

Identification: Similar in shape to the Grey Squirrel (*page 58*), but smaller, slimmer and lighter on its feet. Usually chestnut reddish-brown above (sometimes darker, almost black) with strongly contrasting white belly, while Grey Squirrel is grey or grey-brown above, making less contrast with whitish belly. Any chestnut tinge on a Grey Squirrel is always patchy. In winter, individual Red Squirrels may acquire a greyer hue, but always look uniform, not peppered with white. In winter the ears are very obviously tufted (never in Grey Squirrel). After moult in spring, the tail is the same colour as the back, but bleaches whitish/cream in summer, and has a wispy appearance.

Sounds: Makes softer and higher-pitched sounds than Grey Squirrel. The main call is a subdued squeal. Also makes quiet, staccato "*chuck*" calls with a clucking quality.

Signs: Red Squirrel signs are usually indistinguishable from those of Grey Squirrel.
Nests: Drey like a large birds' nest, about 30 cm in diameter and containing leaves, twigs and moss, often built high in the canopy of a spruce (see *page 57*). **Feeding signs:** Gnawed cones and nuts; sometimes it is possible to find tooth-marks on fungi. **Droppings:** Inconspicuous; small (maximum 1 cm long, cylindrical or round), dark, and not clustered. **Tracks:** Same pattern as Grey Squirrel (fore feet inside hind feet), but prints in jump groups are slightly farther apart and heel mark less distinct than in Grey Squirrel. Fore feet with four toes, print 3·0–4·0 cm long × 2·0–2·5 cm wide; hind feet with five toes, print 4·5–6·0 cm × 2·5–3·5 cm.

Habitat: Nowadays the main habitat is coniferous woodland (pine, spruce), in blocks of 50 ha or more. Able to survive in deciduous woodland, although usually outcompeted by Grey Squirrel where both species are present.

Food: Tree seeds, especially of spruce and pine; also acorns, Hazelnuts and Beech mast, berries, fungi, bark, sap. Makes winter stores of nuts, often hidden buried in the ground, re-found by smell.

WHERE TO LOOK/OBSERVATION TIPS

Shyer and harder to see than Grey Squirrel. Often easiest to find during early morning peak in activity. Try scanning the canopy of conifer woods with binoculars, or listen for commotion in the canopy. Often attracted to bird feeding stations, for example in nature reserves and gardens.

▲ In summer, fur thinner than in winter and tail becomes bleached.
▼ In winter, ear-tufts longer and more obvious.

Fungi seemingly more important than for Grey Squirrel, and individuals often store fruiting bodies above ground in autumn.

Habits: Active by day and throughout the year. There is a marked peak of activity in the morning (3–4 hours after sunrise) all year round, and a secondary peak before dusk in summer. Spends the night in a stick nest (drey) or tree hole (den). Solitary, with an individual home range of 1·0– 6·6 ha in deciduous woodland, 9–30 ha in coniferous stands, but animals will gather at nut-feeders and other food-rich sites. Spends more time in the canopy than Grey Squirrel.

Breeding behaviour: May have two seasons (births February–April and July–September), although spring births tend only to occur after a good autumn seed crop. Males and probably females are promiscuous, mating with several members of the opposite sex. Litters of 1–6 kits are raised in a drey, and the young are weaned at ten weeks.

Population and status: Current British population about 140,000, of which at least 75% are in Scotland; 40,000 in Ireland. Common at times until the 1940s, but has suffered long-term decline, partly caused by an outbreak of Squirrel Pox Virus, which invariably kills Red Squirrels. Grey Squirrels, which are more resistant to this virus, have displaced stricken Red Squirrel populations and their presence in mixed populations appears to have a negative impact (they raid Red Squirrel caches and eat unripe nuts before Red Squirrels can get to them).

Compared with Grey Squirrel, the coat is more uniform.

Often builds several dreys and moves between them.

Red Squirrel is lighter and more agile than Grey Squirrel.

Grey Squirrel
Sciurus carolinensis

One of our best-known mammals, the Grey Squirrel is a conspicuous inhabitant of parks, gardens and woodland. Diurnal and bold, it is a frequent visitor to bird feeding stations, sometimes also taking food from the hand. Although agile on the ground, this species' true home is in the trees, where it is remarkably acrobatic, able to run both up and down trunks and to leap from branch to branch. It is not native to Britain or Ireland, but was deliberately introduced from North America between 1876 and 1929. Since then it has spread over much of England, Wales and parts of Scotland. It is implicated in the decline in the native Red Squirrel, and is known to prey upon birds' eggs and chicks and to cause damage to trees and crops.

Introduced; common	
HB:	24–28 cm
T:	19–24 cm
Wt:	400–600 g

Identification: Long, bushy tail, short muzzle and obvious, rounded ears. Larger and considerably bulkier than Red Squirrel (*page 54*), and lacks the characteristic eartufts of that species. Fur predominantly grey peppered with white, but summer and juvenile coats can have distinct but patchy reddish-ginger tinge on the back, flanks and legs. The tail is edged with white. Dark, almost black, colour forms occur sporadically, and have in recent years become commoner in some areas (*e.g.* Bedfordshire, Cambridgeshire, Hertfordshire). In parts of southern England, Grey Squirrel could be mistaken for the superficially similar, but smaller, Edible Dormouse (*page 62*).

Sounds: A hoarse, loud grating call between a sneeze and a growl, often followed by a series of chatters. When moving around in the canopy in summer often makes loud rustling sounds.

Signs: Squirrel signs are plentiful and easy to identify, although it is rarely possible to tell which species is involved. Look for nests (dreys), feeding signs and tracks. **Nests:** Conspicuous dreys, like large birds' nests, roughly spherical, and about 30 cm in diameter; distinguished by the presence of many leaves in addition to twigs; often lined with stripped tree bark, moss *etc*. Typically built more than 6 m above the ground, often very close to a tree trunk where a branch forks, but also among creepers such as Ivy or under thick canopy in the crowns of conifers. Summer nests often no more than a platform of twigs. **Feeding signs:** Chews the scales off a conifer cone to get at the seeds, but always leaves the apex untouched (Crossbills also eat cones but leave more scales, both opened and unopened). Piles of worked cones and scales may build up around tree stumps and favoured feeding areas. Hazelnuts and other nuts are split cleanly in two, but acorns are shredded more untidily. Also strips tree bark, often leaving it hanging in spirals. This may be at any height, but is especially common in the crown; bark may be stripped all around a branch or trunk, known as 'ring-barking'. Often digs holes when burying or searching for nuts (see *page 44*). **Droppings:** Inconspicuous; small (maximum 1 cm long, cylindrical or round), dark and not clustered. **Tracks:** Squirrels leave a unique track pattern, with the smaller fore feet lying inside and slightly behind the hind feet. Digits point forwards, well splayed. Prints of fore feet have four toes and are 3·0–4·0 cm long × 2·0–2·5 cm wide; hind feet with five toes, prints 4·5–6·0 cm × 2·5–3·5 cm wide.

WHERE TO LOOK/OBSERVATION TIPS

Noisy and conspicuous, and much bolder than Red Squirrel. Look for it in suburban gardens with large trees, in deciduous woodland and in parks, even in city centres.

▲ In summer, fur thinner and redder than in winter
▼ In winter, fur dense and grey

Habitat: Woodland of all kinds, but less common in pure coniferous stands than in broadleaf or mixed woods; also parks and gardens. Arboreal, but perfectly at home on the ground, where it spends more time than Red Squirrel.

Food: Mainly vegetarian and largely dependent upon tree seeds such as acorns, Beech mast, chestnuts, hazelnuts and pine cones. Wide range of other food includes bark, flowers, buds, fungi, invertebrates, and birds' eggs and nestlings. Often caches food, burying nuts, mainly in lawns and flower beds, as insurance against bad winter weather.

Habits: Active during the day and throughout the year, although less active in very cold, wet or hot conditions. When displaying aggressive behaviour, the animals stand on their haunches and flick their tails.

Breeding behaviour: Two breeding seasons a year, with births in spring (February–April) and late summer (July–September). In midwinter, rival males are seen racing around in the leafless woodland, chasing a female. Litters are raised in the drey; the average litter size is three (range 1–7) and the young are weaned after 70 days.

Population and status: Current population in UK is estimated to be 2·6 million, with 300,000 in Ireland; it is probably still increasing. There are regular calls for this species to be culled to help protect Red Squirrels, as well as reduce damage to forestry and other interests.

In some areas, black colour morphs occur.

Grey Squirrels leave neatly split nut shells – see also page 44

Pine cones gnawed by a squirrel resemble a chewed corn-on-the cob.

Dreys are built in trees and resemble a bird's nest; they are about 30 cm across and contain many leaves.

Grey Squirrels feed avidly during the autumn and put on weight to see them through the winter.

Edible Dormouse

Glis glis

(Fat Dormouse)

Not native to Britain, this species was introduced to Tring, Hertfordshire, in 1902, when an unknown number escaped from a private animal collection. It is found naturally in Central Europe where the plump hibernating animals have been eaten by humans since at least Roman times, hence its common names.

Introduced; rare	
HB:	13–19 cm
T:	12–15 cm
Wt:	50–250 g

Identification: Resembles Grey Squirrel (*page 58*), with silvery-grey fur and bushy tail, and is almost equally adept at scaling trees. However, it is much smaller, plumper and more delicate, with paler and softer-looking fur than a squirrel's, and much larger eyes. The ears are smaller and more rounded, it has a blunter muzzle and the tail lacks the white fringe often obvious on Grey Squirrels. Common Rats (*page 92*) also climb trees, but are larger and have scaly rather than furry tails.

Sounds: In the breeding season makes a noisy squeak from the canopy, which has a passing resemblance to the begging calls of young Tawny Owls.

Signs: Difficult to spot. **Nests:** In summer, usually nests in a hole in a tree, a cavity in a building or a nestbox, sometimes adding soft lining (*e.g.* paper, moss). In the tree canopy may construct a spherical stick nest similar to a squirrel drey, and sometimes uses dreys and birds' nests as a base. The untidy structure is often lined with fresh green Beech leaves.
Feeding signs: Strips bark (Larch, willows, plums), often leaving hanging spirals. Opened nuts have jagged edges on the shell, rather than the neat chiselling of Hazel Dormouse (*pages 44 & 65*).

Resembles a squirrel, but smaller and with larger eyes; nocturnal, and hibernates in winter.

WHERE TO LOOK/OBSERVATION TIPS

Commonest in the Chilterns, where trips are organised to check boxes. Easy to hear on calm nights in suitable woodland. These rodents frequently invade houses and other buildings (*e.g.* lofts) and can be very easy to see, especially if attracted with apples.

Droppings: Oval-shaped, 10 mm long × 5 mm wide, in large piles of 50+ deposits. Almost any colour. **Tracks:** Fore feet with four visible toes, prints 2 cm long × 2·2 cm wide; hind feet with five visible toes, prints 3 cm × 2·5 cm. Similar to Grey Squirrel but smaller; sometimes leaves similar jump track, with the front paws inside the hind paws.

Habitat: Mature deciduous woodland, especially of Beech, with relatively continuous canopy. Seems to benefit from the presence of conifers and, when adjacent to suitable habitat, freely enters buildings, where it may breed in lofts and roof spaces.

Food: Mainly vegetarian, relying on nuts and fruit, especially apples. Also eats fungi, bark and occasionally insects, eggs and carrion.

Habits: Hibernates from autumn (September, sometimes as late as November) until May, depending on temperature and food supply. For hibernation, selects a hole low in a tree, or underground (*e.g.* among roots or in a Rabbit hole); sometimes uses a building (under floorboards or in wall cavities). When active, strictly nocturnal; sleeps in a nest during the day. Forages mainly in the tree canopy and is an agile and sure-footed climber. Rarely roams farther than 100 m from its nest. A sociable animal that lives in small groups of related individuals, often sleeping together in a nest.

Breeding behaviour: Courtship, with male pursuing female and squeaking, begins in June. There is just one litter a year, in August. The young (average seven in a litter) leave the nest when 30 days old.

Population and status: Native to mainland Europe. British population around 10,000; trend unknown. Although not protected by domestic legislation, the Edible Dormouse is covered by some aspects of EU legislation (see *page 314*).

Hazel Dormouse

Muscardinus avellanarius

Our native dormouse is hard to see in the wild. It spends over half the year hibernating, and when active is strictly nocturnal, living above ground in trees and overgrown hedgerows. Once much commoner, it is now patchily distributed, primarily in the southern half of England and in Wales.

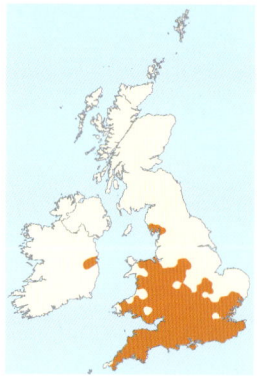

LEGALLY PROTECTED	
Biodiversity List (En, Wa)	
Locally common; introduced Ireland	
HB:	6–9 cm
T:	5·7–6·8 cm
Wt:	15–35 g
	(to 43 g pre-hibernation)

Identification: Very distinctive, with orange-brown fur and bushy tail (compare with tails of mice (*pages 80–91*) or voles (*pages 69–79*)), and small size. The eyes are large and black, and the whiskers are very long; the tail is as long as the body. In the hand (licence required, see *page 312–313*) note the extended length of the toes, an adaptation for climbing. Juveniles and hibernating adults have a greyer tinge to their fur.

Sounds: Not significant. Hibernating animals wheeze as they wake up.

Signs: Look out for feeding signs and nests; rarely leaves tracks or droppings. **Feeding signs:** Discarded hazelnut shells, neatly and smoothly cut with only subtle tooth marks, leaving a round hole in one side (see *opposite*). Shells are often found widely scattered under Hazel. **Nests:** Summer nests are found in a tree hole, on top of a bird's nest or in a squirrel's drey, usually 5–10 m above ground (but sometimes lower down, *e.g.* in brambles). Dormice also use nestboxes if the entrance opens inwards towards the tree. The typical nest is a compact, woven ball of soft vegetation about 10 cm in diameter with no obvious entrance, very often with a lining of stripped Honeysuckle bark. Breeding nests are similar but larger, some 15 cm in diameter. Hibernation nests have thicker walls and are tightly woven enough to be waterproof; they are constructed on or close to the ground, often among roots.

Feeds on fruits, berries and nuts in the autumn, sometimes during the day.

The best way to see a Hazel Dormouse is to join an organised nestbox inspection with a local mammal group. Sometimes visits nut feeders at night. If an animal is flushed from a nest by day, it may return after a few minutes. Very occasionally seen in the canopy, especially in autumn.

Habitat: Typical habitat is deciduous woodland with an understorey, but it is also found in overgrown hedgerows, scrub and, unusually, in conifer plantations. Usually prefers large (>20 ha) blocks of ancient woodland, or a rich, unbroken network of woods and hedgerows, including coppices: important plants include oaks, Sweet Chestnut, Hazel, brambles and Honeysuckle. Occasionally occupies unusual habitats (*e.g.* coastal scrub).

Food: Mainly vegetarian, with flowers and pollen in summer, and fruits, berries and nuts in autumn. Hazelnuts are very important prior to hibernation. Also takes insects in summer.

Habits: Remarkably long hibernation lasts from October to May, when the animal resides in a nest on the ground. In summer it is active from just after sunset until about an hour before sunrise, but in the autumn forages all night when fattening up for hibernation. In the course of a night, may move about 250 m; each home range will have several favoured areas for feeding. Sometimes shares nests, but males at least are territorial during the breeding season.

Breeding behaviour: The breeding season is late, with young usually not appearing until the end of July and into August. Generally one litter per year. Occasionally has a second litter in September, although the young usually fail to survive. The 4–5 young begin to leave the nest after 30 days and may be independent at 40 days, but some remain with their mother for up to eight weeks. Males take no part in raising young and are probably polygamous.

Population and status: Once very common and widespread but now much reduced; population about 45,000. Recently, efforts have been made to slow the decline by reconnecting fragmented habitats to assist dispersal. A National Dormouse Monitoring Programme has been running since 1989 (see *page 318*). Introduced to Ireland (discovered 2012).

Hazelnuts chewed by Hazel Dormouse have a small, smooth-edged round hole on one side – see also page 44

Hazel Dormouse is highly arboreal and nocturnal.

Hibernates in a nest on the ground.

The identification of small mammals

Small mammals can be grouped into five broad 'types'.

DORMICE *pages 62–65*

HAZEL DORMOUSE

DORMICE
- Distinctive bushy tail
- Live above ground in trees and shrubs
- Almost exclusively nocturnal
- Hibernate from late autumn through to spring – unlikely to be caught in standard small mammal live-traps

Hazel Dormouse (*page 64*)

Very small, with a rich, orange-brown coloration. Has a blunt muzzle, relatively big eyes, and small hairy ears.

Edible (Fat) Dormouse (*page 62*)

Squirrel-like and rat-sized with soft, grey fur and large eyes.

RATS *pages 92–97*

COMMON RAT

VOLES *pages 69–79*

BANK VOLE

MICE *pages 80–91*

WOOD MOUSE

SHREWS *pages 100–109*

COMMON SHREW

RATS
- **Larger than voles and mice**
- **Thick, bare tail, with fleshy rings along the length**
- Large ears; prominent eyes
- Mainly nocturnal

Common Rat (*page 92*)

Large, heavy-bodied, with long, greyish-brown fur, a long muzzle. Swims well and often found by water, but also in open countryside and urban areas, away from open water. Can climb.

Ship (Black) Rat (*page 96*)

Larger ears and eyes than Common Rat, with a **longer, thinner tail**. Fur often silky black, but can be brown. Agile; more likely to be seen climbing trees and bushes than Common Rat. Very rare.

ears prominent   large eyes

long, pointed muzzle

COMMON RAT

ears almost hidden   small eyes

short, blunt muzzle

WATER VOLE

VOLES
- **Smaller ears, eyes and shorter tail than mice**
- Tend to run rather than hop
- Active day and night

Field Vole (*page 72*)

Greyish-brown, lacking warm colours. Ears barely protrude from the fur. Shorter tail (50% of body length), and **smaller eyes than Bank Vole**. Found mainly in open grassland.

Bank Vole (*page 69*)

Usually a rich, warm brown colour on back. More obvious ears and eyes and longer tail (66% of body length) than Field Vole. Found mainly in woods and hedgerows.

Water Vole (*page 76*)

Largest vole, approaching the size of Common Rat, told by small eyes and ears almost hidden in the long fur, blunt muzzle and short, furry tail. Often seen swimming.

Orkney (Common) Vole (*page 74*)

Orkney only – Larger than Field Vole, with more prominent, less hairy, ears, paler fur and shorter hairs, giving a smoother appearance.

MICE
- **More prominent ears, eyes and longer tail than voles**
- Often hop as well as run
- Mainly nocturnal

Wood Mouse (*page 82*)

Dark brown above, and greyish-white below; may have an isolated yellowish chest patch – always smaller than on the similar Yellow-necked Mouse. Ubiquitous.

Harvest Mouse (*page 80*)

Very small with a blunt muzzle, small ears and vole-like, prominent beady eyes. **Fur is rich orange-brown above and distinctly white on the belly.** Prehensile tail used for climbing.

House Mouse (*page 90*)

Longer, narrower, more pointed muzzle and smaller eyes than Wood or Yellow-necked Mouse. Fur cold greyish-brown, only slightly paler below. Found mainly in buildings around farms.

Yellow-necked Mouse (*page 88*)

Distinguished from the slightly smaller Wood Mouse by orange-brown upperparts with a more contrasting white belly, larger ears and more prominent, bulging eyes.
The yellow chest patch is broader than deep and usually forms a complete collar. Locally distributed, found mainly in older deciduous woods. Aggressive when handled.

SHREWS – Not closely related to other small mammals
- **Long snouts, tiny eyes, small ears**
- Short, silky fur
- Active day and night

TO PAGE 68

SHREWS *pages 100–109* from page 66

Long snouts, tiny eyes, small ears | Short, silky fur | Active day and night

COMMON SHREW

LONGER-TAILED SHREWS WITH REDDISH TOOTH PIGMENTATION
(see individual species accounts for detailed dentition information)

Common Shrew (*page 100*)

Commonest British shrew, **absent from Ireland**.

Dark brown above and whitish below, with an intermediate band of pale brown on the flanks; 5 unicuspid teeth (U3<U2).

The tail is relatively short compared with that of the similar Pygmy Shrew – around 50% of the length of the head and body.

Pygmy Shrew (*page 102*)

Very small; 5 unicuspid teeth (U3>U2).

Usually shows only two colour shades, brown on upperside and dirty grey on underside, without an intermediate flank band.

Tail comparatively longer (70–80% of the length of the head and body) and thicker than that of Common Shrew.

Water Shrew (*page 104*)

Sharply contrasting black above and white below, and usually a small white spot by each ear; 4 unicuspid teeth.

The feet and underside of the tail are fringed with bristles.

Habitually swims in water.

SHORTER-TAILED SHREWS WITHOUT TOOTH PIGMENTATION

The two white-toothed shrew species are easily distinguished from other shrews by the lack of tooth pigmentation, their shorter tail with bristly hairs, and larger ears.

The only reliable morphological distinction between the two white-toothed shrews is the detail of dentition. *However, in the area covered by this book, geographical location is definitive.*

GREATER WHITE-TOOTHED SHREW

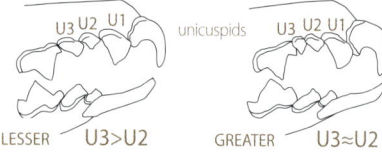

Lesser White-toothed Shrew (*page 106*)

Isles of Scilly (also Jersey and Sark)

Greater White-toothed Shrew (*page 108*)

Ireland (also Guernsey, Alderney and Herm)

U3 U2 U1   unicuspids   U3 U2 U1

LESSER   U3>U2   GREATER   U3≈U2

WHITE-TOOTHED SHREW DENTITION

Bank Vole　　　　　　　　　*Myodes (Clethrionomys) glareolus*

Typically a vole of woodland and scrub, in contrast to the long grass habitats of Field Vole. An abundant species, active by day and night, it is easier to see than most other small mammals, sometimes feeding out in the open. The islands of Skomer in Wales, and Mull and Raasay in the Inner Hebrides, have their own subspecies (as does Jersey in the Channel Islands). The thriving Irish population stems from accidental introductions in the 1920s of animals from Germany.

Very common; introduced to Ireland	
HB:	8–12 cm
T:	3·3–4·8 cm
Wt:	14–40 g

Identification: Mouse-like, but with the small eyes and ears and comparatively short tail typical of a vole. Warm brown or red-brown colour on back distinguishes typical individuals from similar species, but the young are plainer; some are dark, chestnut or grey. Greyish on the underside, with an obvious demarcation from browner fur above. Field Vole (*page 72*) is larger, shorter-tailed (tail length half the body length; two-thirds in Bank Vole), paler brown and its hairy ears are almost hidden in the fur. Skomer Voles tend to be brighter reddish-brown on the upperparts and creamy below.

Sounds: Insignificant squeaks.

Signs: Only runways/nests and feeding signs are usually seen. **Runways/Nests:** Constructs a network of runways and grass nests just below ground, often exposed by turning over logs or other material. The nest is made of moss, leaves, grass and feathers; 2–10 cm below the ground, in soil or leaf-litter, often in dense cover. **Feeding signs:** Nuts are opened neatly, with only the inner rim showing tooth marks (*page 70*), unlike Wood Mouse (see *page 85*). Also eats the flesh of rose hips, and nibbles bark. **Droppings:** Small (smaller than Wood Mouse droppings), brown to black (green in Field Vole); 0·8 cm long × 0·4 cm wide, rounded in section, in latrines off main underground runs. **Tracks:** Very small, even smaller than Wood Mouse tracks. Typical small rodent arrangement of four toes on the fore feet, and five on the hind feet. Prints of fore feet 1·1 cm long × 1·3 cm wide; hind feet 1·7 cm × 1·5 cm.

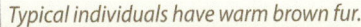

Typical individuals have warm brown fur.

May be seen at bird feeding stations, including those in gardens. Can be encountered by chance scurrying across tracks, or even feeding quietly in the open. Failing this, easily caught in small mammal live-traps (see *page 35*). Most abundant in October–November after the breeding season.

Habitat: Mainly found in woodland, no matter what size, and commonest at the edge of stands. Also occurs in hedgerows, in thick cover and on marshy ground. Usually replaced in open fields by the Field Vole.

Food: Mainly eats buds, seeds, leaves and fruit. Occasionally takes a few small insects. Seeds may be hoarded for the winter, often hidden in the walls of tunnels. In common with many small mammals, the Bank Vole transports seeds in its cheek pouches.

Hazelnuts chewed by Bank Vole are opened neatly and the inner rim shows tooth marks – see also page 44

Habits: Active all year, day and night. Very agile, and will forage high above ground in trees and shrubs. On the ground, it walks and runs, often in quick start-stop dashes, but does not hop. Burrows actively into the ground. Solitary, both males and females having their own home ranges. Ranges overlap in males, but are exclusive for females in the breeding season.

Breeding behaviour: Breeds March–October, occasionally through the winter. There is no pair bond, and females tend to mate with dominant males in the local population. An adult female may produce a litter a month, and as many as eight or nine in the course of a year. Gestation lasts 18–20 days; litter size is usually four. The young are born blind and naked but can run about when 15 days old, are weaned at 18 days, and are independent at 21 days.

Population and status: There are estimated to be 23 million Bank Voles in Britain at the beginning of each breeding season. In the north, the populations fluctuate every 3–5 years. They are found over much of the southern half of Ireland, having spread from the south-west, although the population size is not presently known. The subspecies on Skomer, Pembrokeshire (*Myodes glareolus skomerensis*), Mull (*M. g. alstoni*) and Raasay (*M. g. erica*) are confined to their respective islands, with 7,000 animals on Skomer alone.

Skomer Vole is slightly larger than mainland voles, and often has a reddish back and creamy belly.

▲ Dark individual

▼ Chestnut individual   ▲ Greyish individual (compare with Field Vole, page 72)

See also: Tracks *p. 46*

Field Vole

Microtus agrestis

(Short-tailed Vole)

Although one of Britain's most abundant mammals, the Field Vole is not easy to see in the wild. Faced with many predators, it hides in dense, long grass. Despite this, it is often seen being carried in the talons of a bird of prey, or the mouth of a Fox. Its presence is often given away by a system of runways and nests in the root tangle at the base of grass tussocks.

Very common	
HB:	9–11 cm
T:	2–5 cm
Wt:	20–40 g

Identification: Similar to the Bank Vole (*page 69*), but always grey-brown, lacking warm, chestnut hues. The Field Vole has a shorter tail (tail length half the body length; two-thirds in Bank Vole), and relatively short ears which are hairy on the inside and inconspicuous in its long fur (the ears of Bank Vole are more prominent).

Sounds: Insignificant squeaks.

Signs: Easy to recognise, especially runways in long grass and feeding signs. Runways: A system of runways, tunnels and nests, often obvious under debris on the ground. The tunnels are in the litter at the base of grass tussocks, or just below the surface. **Nests:** Untidy 10 cm diameter balls of shredded grass leaves, built at base of tussocks, in tunnels or under wood and other debris. **Feeding signs:** Cut clippings of grass, often left in runways.

Hairy ears that are hardly noticeable in the fur, and tail less than half the length of the body are the key identification features.

Difficult to see in the wild, but very easy to catch in small mammal live-traps (see *page 35*) set in suitable habitat.

Droppings: Often found within runways, sometimes intermixed with grass clippings. Green, oval: 6–7 mm long × 2–3 mm wide. **Tracks:** Not easily distinguishable from those of other rodents: very small, with four toes on fore feet and five on hind feet. Print of fore feet 1·2 cm long × 1·5 cm wide; hind feet 1·8 cm × 1·5 cm.

Habitat: An inhabitant of long grass, especially ungrazed, but also found in the damp margins of marshland. Can be quite common in young plantations with lots of grass, but less frequent in open woodland, moorland, bogs and hedgerows. Occurs up to 1,300 m in Scotland.

Food: The main diet is grass roots and shoots.

Habits: Active all year and at all times of day, although there may be a peak in activity at dawn and dusk, and during rain. More nocturnal in hot summer weather and, conversely, more diurnal in winter. Both sexes hold small territories in the breeding season, each with core areas of activity; there is more overlap in winter.

Breeding behaviour: Breeding season runs from March–April to September–October, with a sequence of litters, sometimes as many as seven. Young in litter 1–8, average five. Both sexes are promiscuous; dominant males mate with the most females. The young are weaned after 14–21 days, and may reproduce in the season of their birth.

Population and status: The British population is estimated at 75 million animals at the end of the breeding season. However, populations are often cyclical, over a period of approximately four years. Absent from Ireland.

Field Vole runways can be very obvious.

Orkney Vole

Microtus arvalis

(Common Vole)

Abundant in mainland Europe, where it is known as the Common Vole, but absent from Britain except for the Orkney Islands (also present on Guernsey in the Channel Islands, where it is called the Guernsey Vole). It is almost certainly an introduced species, albeit from long ago – remains from Orkney date back 5,400 years. Although recent studies suggest it originated in Belgium, it is considered a distinct subspecies *Microtus arvalis orcadensis*.

Introduced (prehistoric); locally common	
HB:	8–12 cm
T:	2–4 cm
Wt:	19–52 g

Identification: As the only vole on Orkney, identification is geographically straightforward. It is larger than Field Vole (*page 72*) and a paler greyish-brown. The fur is also shorter, giving a smoother, less shaggy appearance, and the ears are almost naked inside, with just a dense fringe of hairs towards the top.

Sounds: Insignificant squeaks.

Signs: Signs include tunnels, nests, grass clippings and droppings
Tunnels: A network of circular tunnels, 3–4 cm diameter, similar to Field Vole, through vegetation at ground level or just below.
Nest: Below ground; similar to that of other voles, made from grass and stalks.
Feeding signs: Small piles of clipped grass (as if cut with scissors), 2–6 cm long, often placed close to a tunnel entrance.

The Orkney Vole is a distinct subspecies of the Common Vole that occurs across Eurasia, and is the only vole found on Orkney.

WHERE TO LOOK/OBSERVATION TIPS
Occurs on the Orkney mainland, and the islands of Westray, Sanday, Rousay, Burray, South Ronaldsay and Eday. May be observed scuttling across roads, but best seen by live-trapping (see *page 35*).

Droppings: Green or black, cylindrical with rounded ends, 6–7 mm long × 2–3 mm wide; often in latrines on path leading to burrow. **Tracks:** Similar to those of other rodents: very small, with four toes on fore feet and five on hind feet. Print of fore feet as Field Vole: 1·2 cm long × 1·5 cm wide; hind feet 1·8 cm × 1·6 cm.

Habitat: Most abundant in unimproved grassland, especially close to moorland. Also found in wet areas with a high water table, such as peat cuttings. Tends to avoid crops and other intensively managed habitats, but occurs in ditches running through agricultural land.

Food: Has a broader diet than Field Vole, encompassing not just grasses, but also a wide variety of herbs. Eats all parts of the plant, including the stems, leaves and roots.

Habits: In common with other voles, active both during the day and at night, with alternating periods of activity and rest lasting about three hours each. Does not hibernate; in winter tends to be more diurnal. Highest densities are recorded from old peat cuttings (over 500 voles/ha), and rough grassland can support over 200 voles/ha. The breeding home ranges are 300–3,700 square metres in extent, and those of different individuals overlap in winter.

Breeding behaviour: Breeds from March–November, with a peak in April–August. Has a variable mating system, with some monogamous pairs in exclusive territories, while some polygamous males hold a territory encompassing those of several females. There are 2–4 litters per year, with 1–6 young per litter (average three), weaned at 20 days.

Population and status: The British population is estimated at 1 million; this is thought to be fewer than a few decades ago, probably as a result of intensification of land management.

Water Vole

Arvicola amphibius

Few animals have declined more steeply in Britain than the Water Vole. Once common and widespread, numbers fell by 94% between 1990 and 1999; it is now almost entirely confined to areas managed for wildlife. Semi-aquatic, Water Voles live in a burrow system typically dug in a soft riverbank, but they are usually seen swimming. A non-amphibious population occurs on some Scottish islands.

LEGALLY PROTECTED (UK)	
Biodiversity List (En, Wa, Sc)	
Fairly common	
HB:	14–22 cm
T:	9–14 cm
Wt:	150–300 g

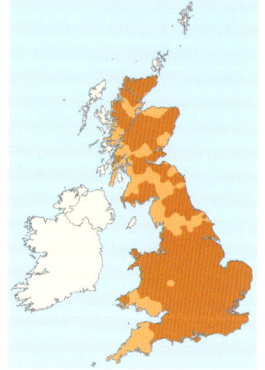

Identification: A large vole, only a little smaller than the Common Rat (*page 92*), which also thrives by water and can swim. Its popular name 'Water Rat' adds to the confusion. Water Vole has a more rounded muzzle and less protruding head than Common Rat, and the ears are furry and not as prominent, almost hidden in the body fur, the colour of which ranges from light brown to almost black (in Scottish populations in particular). It has a short, furry tail, about half the length of the body, unlike the Common Rat's longer, bare tail. It swims buoyantly, albeit with an impression of effort.

Sounds: Makes a distinct plop when it jumps into water. This sound is thought to be intentional, used as a territorial warning. Also makes sharp "*tip*" call, quite similar to the "*kik*" calls of Water Rail, which often occur in the same habitat.

One of the easiest British rodents to find in the wild: a visit to a known site should be successful if you are patient. Most often seen swimming on the water surface, either along the shore or across the main flow of a river or stream, but can also sometimes be spotted sitting on its haunches munching vegetation held in the fore feet.

Signs: Presence quite easy to detect, with burrows, droppings and food leftovers all readily identified. **Burrows:** Holes (4–8 cm diameter) in riverbank at or just above water level. **Feeding signs:** Leaves piles of grass or reed blade clippings, approximately 10 cm long. **Droppings:** Smoothly cylindrical with rounded ends, 8–12 mm long × 3–4 mm wide, and often green. Typically left copiously in latrines for territorial marking purposes, in such places as prominent rocks, raised banks or entry points into the water. **Tracks:** The outermost digits splay to 90°. Fore feet tracks with four digital pads; hind feet with five (six in Common Rat). Prints of fore feet 1·8 cm long × 2·3 cm wide; hind feet 3 cm × 3·1 cm. Stride is 100–120 mm. **Other Signs:** Small muddy patches are created where animals frequently enter the water, often overhung with vegetation.

Habitat: In most of Britain, lives along slow-flowing rivers and in freshwater marshes, rarely straying far from the water's edge. On rivers, favours steep, densely vegetated banks, where it is easy to dig holes that give ready access to the water. In marshes may build its nest above ground at the base of reeds. Also found in the uplands, along streams on peaty soils. On some Scottish islands and parts of the Scottish mainland, it is not aquatic, instead living a mole-like underground existence in dry grassland.

The Water Vole varies in colour from light brown (OPPOSITE) *through dark reddish-brown* (BELOW) *to almost black (particularly in Scotland).*

Food: Vegetarian, eating primarily grasses (including reeds) and sedges as well as a wealth of other waterside vegetation; 227 species of plant have been identified in the diet.

Habits: Active throughout the year and at any time of day, although spends less time outside the burrow in winter. In common with other voles, has alternate spells (2–4 hours) of activity and rest. Where it coexists with Common Rats, it is more active in daylight, while rats are active at night. Water Voles hold linear territories along the riverbank (25–200 m), but concentrate activity in a small area for days at a time. Males, especially, sometimes fight over territory.

Breeding behaviour: The long breeding season begins with mating in March, the first young born in April, and the last litters in September. The mating arrangements vary, with monogamous pairs found in small, isolated colonies, while in larger populations both sexes are regularly promiscuous, many litters sired by two or three males. There are usually six young per litter, born in a nest inside the burrow. They begin to venture from the nest as early as 14 days, finally departing a week later. They may then disperse away from the colony, even overland, with females travelling farther (up to 2 km). Females may have up to five litters a year. The young do not breed until the year after their birth.

Population and status: The British population is currently estimated at 875,000 animals, but is continuing to decline. Habitats have been destroyed or degraded by development, agricultural intensification and pollution, while introduced American Mink (*page 214*), which are major predators, have caused a number of local extinctions. Absent from Ireland.

Burrows just above the water surface are one of the most obvious signs that Water Voles are present. Rats may have similar burrows, but these tend to have wider entrances.

▲ The Water Vole is strictly vegetarian, feeding on a range of plants.
▼ Swims high in the water.

Compare with swimming Common Rat on pages 67 and 95

Harvest Mouse

Micromys minutus

The Harvest Mouse is the smallest British rodent. In summer it lives in the 'stalk zone', climbing acrobatically among the stems of tall vegetation, using its prehensile tail as a fifth limb for grasping. It is well known for building intricate ball-shaped nests woven into the stems, often well above ground.

Biodiversity List (En, Wa)	
Uncommon	
HB:	5–8 cm
T:	5–8 cm
Wt:	5–11 g

Identification: Minute, with a body no longer than a human thumb. The fur on the upperside is generally orange-brown and contrasts sharply with the white underside; however, newly independent juveniles are more greyish-brown. The ears are hairy and relatively small and vole-like, and the nose is blunter than in other mice. It also has rather beady eyes – again, more reminiscent of a vole than a mouse. The sparsely hairy, prehensile tail enables it to climb with ease among the stalks of tall vegetation.

Sounds: Insignificant in the field (makes some ultrasounds).

Signs: Renowned for its ball-shaped nests. Droppings are also identifiable. **Summer nest:** Spherical, made from grass blades; 8–10 cm in diameter (much smaller than a bird's nest). Typically 30–40 cm above ground, woven firmly among vertical stems, with an obvious entrance hole in the side. Usually among grass, but also found among brambles and scrub, sometimes at a height of 1 m or more. Breeding females build several nests in a season, which are green at first, but fade to straw-coloured with time. Nests are well camouflaged in summer, but more obvious in dead vegetation from the autumn.

Lives above ground in summer and into autumn, climbing around stems.

Winter nest: Often on the ground, for example in a tussock or bale of hay. Usually less than 5 cm in diameter, spherical but not as neat as summer nest. **Droppings:** Very small, just 3 mm long and typically found within the nest.

Habitat: Occupies a surprisingly broad range of habitats with tall herbage, including rank grassland, rushy meadows, dry reedbeds, crops (especially oats, wheat and legumes), saltmarshes, ditches and bramble patches.

Food: Eats seeds, fruit, berries and invertebrates, and occasionally fungi and mosses. Sometimes catches butterflies or moths.

Habits: Climbs above ground among vegetation. When alarmed may simply drop down to the ground. Mainly nocturnal in summer, with a peak of activity at dusk; more diurnal in winter, when it lives mainly at ground level, often occupying burrows made by other rodents.

Will sometimes feed on grain, as their name suggests.

▲ *Typical individual. Note the blunt nose and small ears, compared with other mice.*
▼ *Harvest Mice forage on the stems and branches of herbs, using their prehensile tail as a fifth limb. This is a relatively dark individual.*

Breeding behaviour: Breeding season May–October, sometimes into December if conditions are mild. Both sexes are promiscuous and the males do not tend the young. Raises a maximum of three litters per year, with 1–8 young in each, in different nests. The young are independent of the mother at 15–16 days, but continue using their natal nest.

Population and status: The British population at the end of the breeding season is estimated at 1,425,000. Once much commoner in crops, but has almost certainly declined due to the effects of combine harvesting and the use of pesticides reducing food availability. Absent from Ireland.

Litters can number up to eight young.

Summer nests are a spherical ball of grasses, usually low down in dense vegetation.

See also: Feeding signs *p. 44* | Tracks *p. 46*

Wood Mouse

Apodemus sylvaticus

(Long-tailed Field Mouse)

One of Britain's most abundant mammals, far outnumbering the better known House Mouse and Common Rat. It occurs in many different habitats and often enters buildings, although causes fewer problems than House Mouse. Often seen in gardens and sheds, or running across rural roads at night.

Very common	
HB:	6·1–10·3 cm
T:	7·1–9·5 cm
Wt:	13–27 g

Identification: Typical mouse shape with a long tail, pointed muzzle and large ears. Dark brown above and greyish-white below, with yellow-brown flanks; the demarcation between the dark and pale fur is not as clear as is usual in Yellow-necked Mouse (*page 88*). Some individuals have a yellowish spot isolated in the middle of the chest (compare with the full collar found on Yellow-necked Mouse, see *page 87*). The prominent eyes and ears distinguish it from the similar-sized House Mouse (*page 90*); it also has larger hind feet and lacks a musky smell. Young animals are grey-brown and very similar to House Mice.

Sounds: Insignificant squeaks, and some ultrasonic calls.

Despite its name, the Wood Mouse is found in many different habitats, besides woodland.

Despite its abundance, not easy to see due to its nocturnal habits. Visits bird tables at night (low platforms in particular will often yield results) and sometimes by day. Often the most numerous capture in small mammal live-traps (see *page 35*). Will occupy bird, bat or dormouse boxes.

Signs: Quite easy to detect from feeding signs, including messy leftovers. **Feeding signs:** In favoured sites (*e.g.* among tree roots, on stumps, or under logs) often leaves food remains, including nut shells, seed coats, fragments of catkins and small snail shells. Caches of uneaten food are stored in chambers inside burrows, but also among roots and in crevices in walls. Cones are chewed to the core, sometimes remaining attached to the tree, the scales very neatly removed. Hazelnuts are sliced near the top, with small but obvious tooth marks on the outer surface of the shell, just below the edge of the hole: Bank Vole leaves tooth marks on the inner rim (see *page 44*). Other nut shells, including acorns and Beech mast, are also neatly gnawed, and chewed Horse Chestnut conkers show tooth marks on the white flesh. Seeds from the pips of rose hips are also taken, but the flesh is left (the opposite in Bank Vole, *page 69*). **Droppings:** Larger than Bank Vole droppings, up to 8 mm (usually 3–5 mm) long, thick and rounded in section, and usually very dark.
Tunnels/Nests: Lives in burrow system below ground, with round entrances often under tree roots. **Tracks:** Very small, with the typical arrangement of a small rodent: four toes on the fore feet and five on the hind feet. Fore feet prints 1·3 cm long × 1·5 cm wide; hind feet 2·2 cm × 1·8 cm.

Habitat: Abundant in woodland, but occurs in many other habitats: gardens, road verges, hedgerows, arable fields, heathland and moorland, the edge of wetlands and in dry stone walls; found on most small islands.

Food: Has a varied diet: seeds, nuts, buds, berries, stems, fungi, moss and galls, and a range of animal food including snails, insects, centipedes and worms. Acorns, Sycamore seeds, hazelnuts, blackberries, wild oats, wheat and sweetcorn are especially favoured.

An isolated yellowish spot in the middle of the chest can be seen clearly in this individual.

Nuts and hips chewed by Wood and Yellow-necked Mice both show a small hole with obvious tooth marks – see also page 44.

Habits: Active all year round; primarily nocturnal, with peaks of activity at dawn and dusk, and sometimes active by day in the summer. Inhibited by cold and wet conditions, and strong moonlight. May travel up to 1,200 m at night on the ground; also climbs above ground in trees and bushes, although rarely above 5 m. Lives in a burrow system that persists through generations, often with many entrances and chambers. In winter, individuals may share burrows, but in the breeding season they are more territorial. Moves around by running, but can also hop and jump, especially to avoid a predator.

Breeding behaviour: Breeding season at least March–October, sometimes into winter; peak is July–August. Females may produce four litters a year, occasionally more, with 4–7 young per litter. The young are weaned at about 18 days, and females can breed in the season of their birth. Both sexes are promiscuous; males have overlapping home ranges, encompassing several females' territories.

Population and status: A pre-breeding population of 38 million grows to an estimated 114 million in autumn, when most small mammal populations are at their peak. The Irish population is estimated at ten million.

Yellow-necked Mouse (ABOVE) has a more orange-brown back, whiter underparts, larger ears and more prominent bulging eyes than Wood Mouse (BELOW).

Wood Mouse and Yellow-necked Mouse compared

	Wood Mouse	Yellow-necked Mouse (*page 88*)
UPPERPARTS:	Dark brown	Orange-brown
UNDERPARTS:	Greyish-white; less contrast with upperparts than Yellow-necked	white; more contrast with upperparts than Wood
CHEST PATCH:	May have an isolated yellowish chest patch – always smaller than on Yellow-necked (akin to a neck-tie)	**Yellow chest patch is broader than deep and usually forms a complete collar (akin to a 'V'-necked sweater)**
OTHER FEATURES:	Yellow necked Mouse is slightly larger, with larger ears and more prominent, bulging eyes	

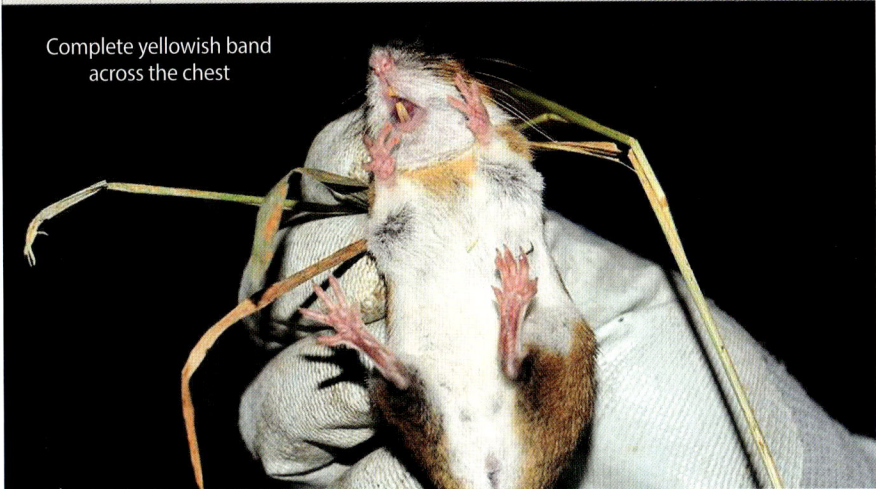

Complete yellowish band across the chest

▲ Yellow-necked Mouse
▼ Wood Mouse

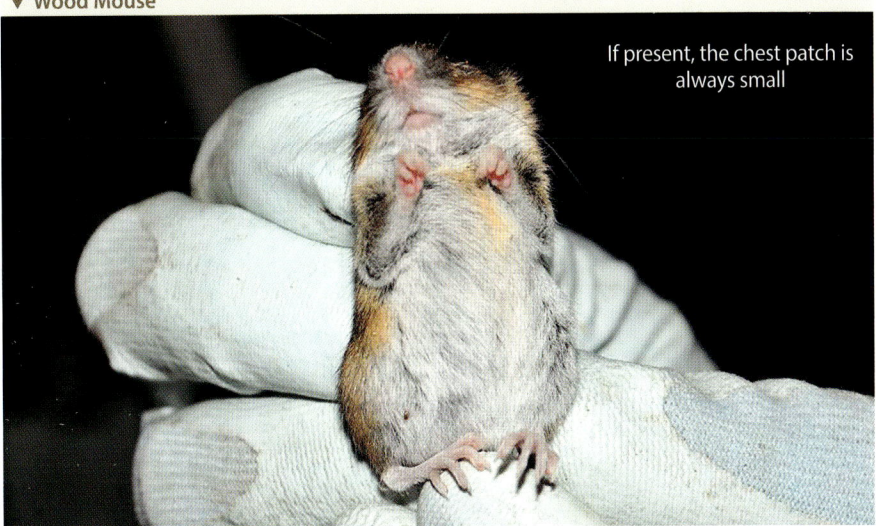

If present, the chest patch is always small

See also: Feeding signs *p. 44* | Tracks *p. 46*

Yellow-necked Mouse

Apodemus flavicollis

Much less common than Wood Mouse, the larger Yellow-necked Mouse occurs mainly in southern England and is strongly associated with mature deciduous woodland.

Identification: Similar to Wood Mouse (*page 84*) but larger and more contrasting in colour – rich orange-brown above with a brighter white underside. Diagnostically, it has a complete yellowish collar across the chest; Wood Mouse can have a variable patch of yellowish or chestnut-tinged fur in the middle of the chest, but never a continuous band (see *page 87*). Yellow-necked Mouse also has larger ears, a longer, thicker tail and more bulging eyes, and a more aggressive nature.

Sounds: Various insignificant squeaks.

Signs: Include nests, nut shells, seed remains and droppings, although none are distinguishable from those of Wood Mouse (see *pages 42–46*). **Tracks:** Slightly larger than those of Wood Mouse. Prints of fore feet 1·6 cm long × 1·8 cm wide; hind feet 2·4 cm long × 1·9 cm wide.

Habitat: Inhabits long-established, mature deciduous woodland, especially with abundant dead wood and a rich understorey, where a supply of tree seeds is guaranteed. Will move into nearby gardens and hedgerows and sometimes enters rural buildings (*e.g.* roof spaces), especially in winter.

Uncommon	
HB:	9·5–12 cm
T:	8–11 cm
Wt:	14–45 g

Yellow-necked Mouse is best told from Wood Mouse by a complete yellowish band across the chest – see page 87.

Usually seen only by live-trapping (see *page 35*), when much more aggressive than Wood Mouse, jumping about wildly, squealing and urinating, and attempting to bite the handler. Even before close examination, you can be relatively certain of an animal's identity by this behaviour alone.

Food: Tree seeds (*e.g.* Beech mast) form much of diet. Also takes buds, fruit and some insects and their larvae.

Habits: Active all year, but entirely nocturnal. Lives in a burrow system down to about 50 cm depth, with several nests, and often stores food underground. Spends much time foraging in trees, recorded up to 23 m from the ground. Home ranges cover about 0·4–1·5 ha. When evading a predator, can leap almost 1 m into the air.

Breeding behaviour: Breeding season February–October (begins earlier than Wood Mouse), sometimes into winter after a bumper fruiting season. May live as monogamous pairs, and have three litters a year. Litter sizes range from 2–11, with an average of five young per litter. Born blind and naked, their eyes open at 13–16 days, and young born early in the year may themselves breed at 2–3 months old.

Population and status: Population about 750,000, although this is an estimate based on inadequate data. This is a generally uncommon and localized species, but there is no evidence of any decline. Absent from Ireland.

Larger ears and more bulging eyes than Wood Mouse

House Mouse

Mus (musculus) domesticus

The world's most widespread land mammal, apart from humans, the House Mouse probably originated in the Middle East and spread here during the Bronze Age. It has a long history of close association with people, and tends to be found inside or near buildings, where it can be a serious pest.

Introduced (prehistoric); common	
HB:	6–10cm
T:	6–10cm
Wt:	12–22g

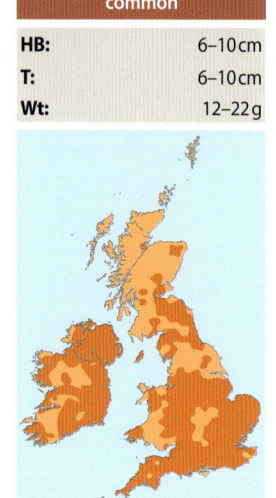

Identification: Greyer than Wood Mouse (*page 84*) and Yellow-necked Mouse (*page 88*), with relatively smaller eyes, ears and hind feet; more pointed muzzle; and less sleek in appearance. The long, ringed tail can look rat-like, although the small size and sharper snout of House Mouse should distinguish it even from young rats.

Sounds: Insignificant squeaks. Gnawing, scratching and running often heard in enclosed spaces.

Signs: Include droppings, urine patches, untidy nests and food leftovers. Often leaves an unpleasant stale smell. **Droppings/urine:** Droppings cylindrical, 6–7 mm long × 2 mm diameter (smaller than Wood Mouse), rounded in section; any colour. Often in piles (up to 4 cm high), mixed with dirt, grease and urine. **Feeding signs:** An erratic forager, moving rapidly from place to place (up to 20 sites a night), eating little and often. Has a distinctive habit of de-husking grain and eating just a proportion, leaving the remnants as 'kibbled grain'.

Greyer, more uniform fur, smaller eyes and ears, and a more pointed muzzle than Wood Mouse

WHERE TO LOOK/OBSERVATION TIPS

Not as easy to see as might be expected, unless you have an infestation. Good places include farming areas (piggeries, hen houses, hay ricks, straw stacks), zoos (in enclosures meant for other species!) and rubbish tips. Often easy to see in the London Underground, scurrying beneath the tracks.

Tracks and runways: Tracks very small, four toes on the fore feet and five on the hind feet (typical for rodents). Prints of fore feet 1 cm long × 1·3 cm wide; hind feet 1·8 cm × 1·8 cm. Has regular pathways in buildings which may be marked by dark, greasy smears, and also by tracks in dust. **Nests:** Untidy balls of shredded paper and fabric, and other soft material gathered from the vicinity of the nest. **Damage:** Chews paper (although less than Wood Mouse), electric wires, plastic *etc*. and can cause considerable damage. Droppings are often found in food stores and result in spoilt grain.

Habitat: Found mainly in buildings; also on waste ground and around livestock. However, also capable of living well away from buildings (*e.g.* in some arable fields and on offshore islands). Not normally found in woods, where it cannot compete with Wood Mouse or Yellow-necked Mouse.

Food: In buildings, will feed on almost anything, although grain is the staple food. Outdoors, invertebrates and wild seeds form a greater proportion of the diet.

Habits: Active all year; nocturnal, with bursts of activity through the night when searching for food. Usually sleeps and grooms during the day. Burrows into the ground or lives in holes (2–3 cm diameter) with passageways of variable length and complexity; in buildings may live in drains, roof spaces, behind walls *etc*. Home ranges vary from tiny (<5 square metres) in grain stores to 100 square metres in fields.

Breeding behaviour: Can breed at any time of year, but outdoor populations only in summer. Each female produces 5–10 litters a year of 5–8 young, weaned at 23 days. Young are sexually mature when just 5–6 weeks old. Often lives in breeding groups with several females served by a single dominant male. Uniquely, these females may pool their litters in communal nests and nurse other females' offspring.

Population and status: Pre-breeding British population tentatively estimated at 5·4 million; no estimate for Ireland.

House Mouse droppings in an attic

See also: Droppings *p.42* | Tracks *p.46*

Common Rat

Rattus norvegicus

(Brown Rat, Norway Rat)

The Common Rat is not native; first recorded around 1720, it probably originated from Central Asia. Rats repulse many people, but their malign influence today is exaggerated. The old adage that "you're never more than six feet away from a rat" is not true: less than 0·5% of dwellings host rats. Nevertheless, rats carry serious diseases and can be found in large numbers, especially in some of the unwholesome places they inhabit.

Introduced; very common	
HB:	21–29 cm
T:	17–23 cm
Wt:	200–400 g

Identification: Ten times the weight and up to three times the length of a House Mouse (*page 90*), with long, greyish-brown fur and a long, scaly, relatively broad tail with fleshy rings along its length. Swims well, and could be confused with a Water Vole, but the tail is much longer, and the rat has a more pointed snout, and larger ears which have fine, rather than thick fur (see *page 67* and compare *page 95* with *page 79*). The rare Ship Rat (*page 96*) has larger eyes and ears (which are almost hairless) and a longer, thinner tail.

Sounds: Most vocalisations are ultrasonic, but also makes squeaks, whistles and tooth-grinding noises, and can often be heard running in enclosed spaces.

Most often associated with human habitation …

WHERE TO LOOK/OBSERVATION TIPS

Food left on the ground attracts rats, so try bird feeding stations, tips, picnic sites, waste places, the edges of arable fields *etc*. Often seen along rivers, canals, and seashores. Quite adept at climbing trees, bushes and hedgerow shrubs.

Signs: Leaves many signs, including runs, burrows and piles of droppings; also causes damage to household items. **Runs:** Worn paths 5–10 cm wide on the ground, usually close to cover, and can be up to 400 m long. When used continuously inside buildings, runs acquire a greasy, black smear.

Droppings: Longer than those of most other rodents, 1·2–2·0 cm long and 0·6 cm in diameter; sausage-shaped, pointed at both ends, and variable in colour; often grouped in latrines. **Burrows:** Digs burrows into the ground, often on the side of a bank. These are of variable size, but typically 6–8 cm in diameter. Burrows usually go down no more than 50 cm, and can be extensive, with many chambers and entrances. Similar to those of Water Vole, but the excavated earth is usually piled in a heap at the tunnel entrance (unless the hole goes vertically down, *e.g.* to a sewer). Common Rats also create burrow systems in stored hay and straw. **Damage:** Will gnaw through electric wiring, plastic *etc.* to get at food, and can cause serious damage. **Tracks:** As with other rodents, four toes on the fore feet, and five on the hind feet. Prints of fore feet 1·8 cm long × 2·5 cm wide; hind feet 3·3 cm × 2·8 cm; long heel area distinctive.

Habitat: Well known for inhabiting sewers, rubbish tips and buildings, but also occurs in 'wilder' places, including hedgerows, saltmarshes, beaches, offshore islands, waterways and field margins. Attracted to food put out for livestock, and to garden bird feeding stations.

… but can be found in a wide range of habitats.

Food: Omnivorous, favouring grain, seeds and root crops, but a wide variety of other food items has been recorded, including fish, molluscs, birds and even soap. Very cautious about novel foods, taking fragments at first to test them out. Also stores food.

Habits: Active all year and mostly nocturnal, with peak activity at dusk and again from 3 am until dawn. However, rats are frequently seen during the day, especially when food is abundant and the population high, or where there is a high risk of nocturnal predation by foxes. Typically forages up to 500 m when seeking food, although has been recorded travelling up to 3·3 km. Nests are built inside the burrows, using dry grass as well as paper, fabrics *etc.* Rats live socially in small colonies, made up of territorial 'clans' (a male with one or several females and their young) which occupy a single burrow system. The clans with the highest-ranking males in the colony live closest to available food, and defend their territory aggressively, especially at high population densities.

Breeding behaviour: Breeds throughout the year, with females producing five litters (mean litter size nine) per year on average. The young are weaned at 21–23 days. Socially dominant male rats may have a harem of females, others only a single mate, but only the female tends the young.

Population and status: British population is probably about ten million, of which 6–7 million are in buildings, farms and other industrial centres. The population in Ireland is unknown.

Common Rats are often sociable and seen out in the open.

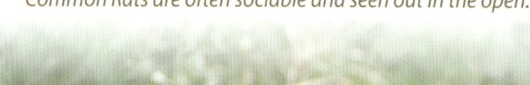

▲ Although wider, Common Rat burrows situated alongside watercourses could be confused with those of the Water Vole.

▼ Common Rat often swims and can be confused with Water Vole. However, it is less buoyant, swimming lower in the water with rear-end raised. Its long, scaly tail, larger ears, more prominent eyes and longer muzzle should confirm identification (compare with swimming Water Vole on pages 67 and 79).

Ship Rat

Rattus rattus

(Black Rat, Roof Rat)

Now very rare in Britain and Ireland, the Ship Rat has declined almost to extinction over the past 300 years, through a combination of targeted control, improved food hygiene, and competition from the Common Rat. A native of the Indian subcontinent, it also is less able to cope with chilly and damp climates than the Common Rat. The Ship Rat inadvertently played a major part in British human history, carrying fleas that were vectors of bubonic plague.

Biodiversity List (Sc)	
Introduced; very rare	
HB:	15–24 cm
T:	11·5–26 cm
Wt:	145–280 g

Identification: Generally smaller than the Common Rat (*page 92*), with sleek fur, larger eyes and ears and a thinner tail. If examined closely, the ears are almost hairless, in contrast to the fine-furred ears of Common Rat. The fur may be black, but colour varies considerably and is not a reliable identification feature. The Ship Rat is more timid than the Common Rat and runs faster.

Sounds: Not distinguishable from Common Rat.

Signs: Very similar to Common Rat; key differences are:
Runs: Greasy smears from well-used runs inside buildings form less-continuous loops around obstructions than those made by Common Rats. **Droppings:** Slightly smaller than those of Common Rat: 1–1·2 cm long and 0·2–0·3 cm in diameter, more round-ended, and sometimes deposited singly. **Tracks:** Smaller than Common Rat, prints of fore feet 1·5 cm long × 2 cm wide; hind feet 2·5 cm × 2·5 cm. Trail left by a running Ship Rat shows a diagnostic heavy, continuous tail drag.

Habitat: On Lambay Island, Co. Dublin (*arrow on map*) found on cliffs and in buildings. Otherwise in ports, especially inside grain stores.

Food: Omnivorous, with a preference for grain and fruit; apparently less partial to meat than Common Rat.

Habits: Active all year and nocturnal, but may emerge during the day if food is in short supply, or if undisturbed. More agile than Common Rat and in much of its worldwide range forages in trees. Occurs in groups with a dominant male and several females, one or two of which are dominant. Females are aggressive when feeding.

Breeding behaviour: Breeds in summer (March–November). A female may have 3–5 litters of about seven young a year. The young are weaned after 20 days and can become sexually mature at 12–16 weeks.

Population and status: Possibly now only on Lambay Island off the coast of north County Dublin in Ireland; probably extinct elsewhere, although small populations may come and go (especially in ports). To conserve seabird colonies, eradicated from Lundy, Devon, in 2005, and recently from the Shiant Islands (Outer Hebrides), where previously common.

The larger Common Rat (ABOVE) has relatively smaller ears and eyes, and a thicker tail than Ship Rat.

Not all Ship Rats are black – size, body shape, tail, ear and eye features are key to identification.

Eurasian Beaver

Castor fiber

The Eurasian Beaver is a British native, once widespread, but hunted to extinction for its pelt and for 'castoreum', an oil extracted from sacs near the base of the tail. This oil was used in perfumery, and believed to cure headaches and other ailments. Beavers were already rare by Saxon times, but possibly survived near Loch Ness, Highland, until the early 16th century. Following a formal reintroduction trial, there is a small population at Knapdale Forest, Argyll. The populations elsewhere have arisen from unauthorised releases.

Reintroduced; rare	
HB:	75–90 cm
T:	28–38 cm
Wt:	12–38 kg

The Argyll and Devon populations relate to trial reintroductions.

Identification: Distinctive on land, but usually seen in the water, where it could be confused with an Otter (*page 210*) or American Mink (*page 214*). Has a large, flat-topped head, small eyes and ears, thickset body and, diagnostically, a scaly, horizontally flattened tail. The hind feet are webbed. Much larger than Water Vole (*page 76*) or Common Rat (*page 92*), which are also found in the same habitat; see also Coypu (*page 264*), which lacks the flattened tail (introduced to, but now extirpated from Britain). Canadian Beaver *Castor canadensis* (see *page 262*) is more commonly kept in captivity in Britain than Eurasian Beaver and individuals occasionally escape. It is very similar to Eurasian Beaver, but has a smaller, more rounded head, a shorter, wider muzzle and a wider tail.

Sounds: Makes a distinctive loud sound by slapping the tail on the water surface when alarmed. Also growls and hisses, but these sounds are rarely heard in the wild.

Beavers have a unique flattened, scaly tail.

WHERE TO LOOK/OBSERVATION TIPS
Guided tours provide an opportunity to see Beavers in the Scottish reintroduction project (scottishbeavers.org.uk). The animals are easiest to see at dawn and dusk.

Signs: Can radically alter the local environment through its activities, and leaves obvious signs. Often uses an inconspicuous burrow as its lodge, with entrances below water level. **Lodge (if not in burrow):** A pile of twigs, branches and earth, up to 2 m high, often built in shallow water. Also makes similar but much smaller scent mounds, used for odour marking. **Dams:** Constructed from logs, branches, rocks and mud; built across flowing water to ensure that the water covers the lodge entrances. Beavers also clear channels through aquatic vegetation. **Feeding signs:** Beavers fell waterside saplings and larger trees by gnawing through the trunk about 50 cm above ground, leaving pointed broken ends. They also strip bark on standing trees, typically leaving visible tooth-marks. **Tracks:** Tracks show diagnostic webs between toes. Prints large: fore feet 5·5 cm long × 4·5 cm wide; hind feet 15 cm long × 10 cm wide. Feet have five digital pads, but usually show only three or four in prints.

Signs of damage by Eurasian Beaver are readily identified, particularly trees that have been gnawed at the base.

Habitat: Largely aquatic, favouring wooded river systems. Usually seen swimming in lakes, rivers and tree-fringed pools, but will graze on bankside vegetation. Favours deciduous woodland, especially with birches, willows, Alder, Aspen.

Food: Vegetarian. In winter relies almost entirely on bark, which is harvested in autumn and stored in the lodge. In spring and summer takes a wide variety of plant material, including aquatic plants, roots, leaves, shoots and grass.

Habits: Typically nocturnal, but in undisturbed places sometimes seen during the day. Active all year, although less so in winter, and may remain in the lodge during freezing conditions. The social unit consists of an adult pair with young from the current year, and also sometimes from the previous year. Each colony occupies a home range with one or more lodges.

Swims low in the water, on a steady course.

Breeding behaviour: Unusually for a rodent, the breeding system is based on a monogamous pair. Mating occurs in mid- to late winter and kits are born at the end of May or in June. Litter size is 2–6 (average three). The young are born fully furred with their eyes open, and can swim within lodge entrances immediately. They are weaned in 2–3 months, but do not leave the colony until over a year old.

Population and status: Three families of Eurasian Beavers were introduced in May 2009 to Knapdale, Argyll, and a small population now lives wild. A 5-year release trial on the River Otter in Devon has been authorised by Natural England. A larger population resulting from releases in the Tay and Earn catchments now numbers more than 150 animals. Further formal trials (*e.g.* in Wales) are planned. Eurasian Beaver is covered by some aspects of EU legislation (see *page 314*).

Common Shrew

Sorex araneus

The commonest British shrew, against which others should be compared, although absent from Ireland. It is typically glimpsed running close to the ground, usually within dense vegetation, its long snout constantly quivering in search of food: it is a voracious hunter of small invertebrates. Although shrews form a major part of the diet of Barn and Tawny Owls, they seem to be distasteful to many other predators since uneaten bodies are often encountered in the wild.

Identification: Typically dark brown above and greyish white below, with a distinct intermediate brownish patch on the flanks. Common Shrew is smaller and less contrasting than Water Shrew (*page 104*) and larger than Pygmy Shrew (*page 102*), but has a proportionally shorter tail (50–60% of length of head and body, as opposed to 65–80%), which is hairy in young animals, particularly at the tip. In the hand (a licence is required deliberately to catch shrews using live-traps, see *pages 35* and *312–313*), identification can be confirmed by dentition (see *opposite*).

Sounds: Makes an audible high-pitched buzzing squeak.

Signs: Relatively few, but droppings and tracks may be found occasionally. **Tracks:** Rarely seen. Small: prints of fore feet 0·8 cm long × 0·9 cm wide; hind feet 1 cm × 1 cm. As with other shrews has five digits, three pointing forwards and two splayed either side; fore and hind prints close together. **Droppings:** Granular, containing hard parts of insect bodies; flaky when dry (rodent scats are hard). Usually black, 3–4 mm in length; hard to find.

LEGALLY PROTECTED (UK)	
Very common	
HB:	5·4–8·7 cm
T:	3·2–5·6 cm
Wt:	6–12 g

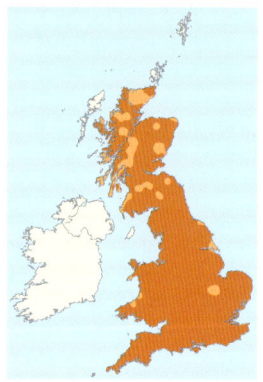

The teeth of all shrews, apart from the white-toothed shrews, are reddish

Typically shows brownish flanks between the dark back and pale belly.

WHERE TO LOOK/OBSERVATION TIPS

Hard to see as hides in thick cover, but try turning over logs or other larger items. Can be tempted into the open by leaving maggots and waiting quietly. In mammal traps, is often caught at dawn.

Habitat: Ubiquitous wherever there is thick vegetation at ground level, in grassy and woodland habitats. Often found in gardens, and particularly abundant along road verges. Mainly forages at ground level, but can climb.

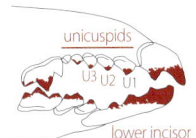

Food: A voracious, non-selective predator of small invertebrates, taking everything from mites to large earthworms, and especially beetles, worms, woodlice, slugs and snails. It needs to consume up to 90% of its body weight every day to sustain its active lifestyle, and catches more prey underground than other shrew species.

*Common Shrew: five unicuspid teeth 3rd (U3) **smaller than** 2nd (U2); lower incisor serrated*

Habits: Can be seen (and heard) at any time of day and night, but with peaks of activity at dawn and dusk. Throughout this time it has alternating 'shifts' between resting and hunting. Lives secretively in leaf-litter, using narrow runs and often the burrows of other small mammals. It is territorial, each individual occupying an area typically 370–630 square metres; shrill squeaks usually indicate aggressive encounters between individuals.

Breeding behaviour: The breeding season lasts from April to September, with first births generally in May. There are typically 4–8 young (up to 11) in a litter, which are born naked and blind, and weaned at 22–25 days. Both sexes are highly promiscuous and the relationship between the male and female lasts only for the act of mating, in which the male holds on to the back of the female's neck with its teeth. Each female produces up to four litters over the summer, but as the year progresses litters become smaller and many adults die. By the winter, most of the population consists of young born the previous summer.

Population and status: The British population is estimated at 42 million, with no definite trend. Absent from Ireland.

Some individuals, mainly immatures, have dark flanks and can be confused with Pygmy Shrew.

See also: Tracks *p. 47*

Pygmy Shrew

Sorex minutus

The smallest terrestrial mammal in Britain and Ireland, the Pygmy Shrew lives in thick vegetation at ground level, where there is plenty of invertebrate food. It is very widely distributed and occurs in some extreme habitats, including blanket bogs and mountain tops, where other shrews do not occur.

LEGALLY PROTECTED	
Common; probably introduced to Ireland	
HB:	4·0–6·4 cm
T:	3·0–4·6 cm
Wt:	2·5–7·5 g

Identification: Markedly smaller than Common Shrew (*page 100*), paler brown and greyish-white below, lacking the intermediate brown flank colour, although juvenile Common Shrew often lacks this colour too. The tail is relatively long (65–80% of head and body), hairy and often tufted at tip in young animals, and thick at the base; it is usually more obviously dark above and pale below than in Common Shrew. The snout is more conical and the head more domed than any other shrew. In the hand (a licence is required deliberately to catch shrews, see *pages 35* and *312–313*), identification can be confirmed by dentition (see *opposite*).

Sounds: High-pitched squeaks, although these are above the hearing range of many people. A bat detector (see *page 134*), set at 40 Hz, is the most reliable way to hear them.

Signs: All signs of this animal are small and obscure, and probably indistinguishable from those of Common Shrew. **Tracks:** As Common Shrew.

Habitat: Open areas with rich ground cover, including heathland, moorland, grassland and marshy areas – it frequents damper places and higher ground than Common Shrew. Sometimes common in deciduous woodland. Spends less time underground than Common Shrew.

Tail relatively longer and fatter than that of Common Shrew

WHERE TO LOOK/OBSERVATION TIPS

Not easy to find: less common than Common Shrew, occurring at lower densities, although is more diurnal. Reluctant to enter small mammal traps (see *page 35*), and so small that it does not always trip the catching mechanism. More frequent in mountains, moorlands and bogs than Common Shrew, and sometimes glimpsed on dry stone walls. Sometimes found in dormouse boxes.

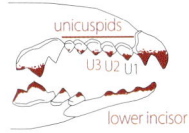

Pygmy Shrew: five unicuspid teeth 3rd (U3) ***larger than*** *2nd (U2); lower incisor serrated*

Food: A wide range of invertebrates, including beetles, spiders, harvestmen and woodlice (preferred size 2–6 mm), but not earthworms – it hunts on, rather than under, the ground.

Habits: Active throughout the day and night. Does not hibernate: most adults die before winter. Although it is territorial, as populations are sparser than those of Common Shrew, it is less frequently involved in aggressive interactions. Follows the runs and inhabits the burrows of other small mammals but is unable to dig.

Breeding behaviour: Often nests under logs and rocks. The breeding season lasts from April to October, with the first young born in May and births peaking in June. Females may have two or more litters of 4–6 young per year. Both sexes are promiscuous and males take no part in rearing the young. The young are weaned at 22–25 days, and first breed in the year following their birth.

Population and status: Estimated to be around 8·6 million at the beginning of the breeding season, with no clear trend. Probably introduced to Ireland by early human settlers; no population estimate.

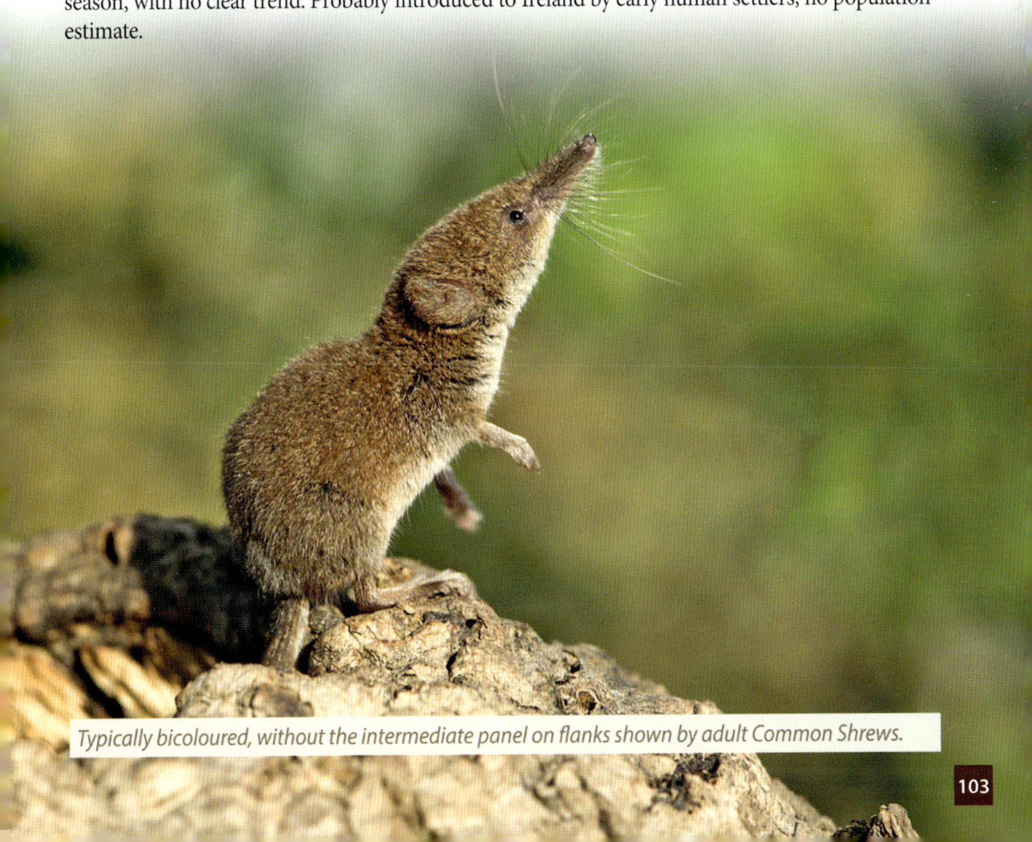

Typically bicoloured, without the intermediate panel on flanks shown by adult Common Shrews.

Water Shrew

Neomys fodiens

(Eurasian Water Shrew)

The elusive Water Shrew is equally adept on land and in the water, where it can swim and dive below the surface. A fringe of hairs on each hind foot aids swimming, and two rows of stiff hairs on the tail act like a keel. It has a poisonous bite, which can be painful to humans, and lethal to other small mammals.

LEGALLY PROTECTED (UK)	
Uncommon	
HB:	6·3–9·6 cm
T:	4·7–8·2 cm
Wt:	8–23 g

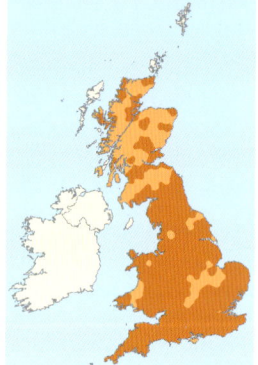

Identification: Similar to Common Shrew (*page 100*), but slightly larger and much darker above, in stark contrast to the white belly (although some individuals can be entirely black); this sharp colour divide continues along the tail. Most Water Shrews show a small white ear tuft (some Common Shrews may also show a hint of this feature). Other British shrews do not usually swim, and do not forage underwater. It is highly buoyant and therefore has to leap to dive below the water surface. In the hand (a licence is required deliberately to catch shrews using live-traps, see *pages 35* and *312–313*), identification can be confirmed by dentition (see *opposite*).

Sounds: Relatively noisy for a shrew, producing shrill squeaks.

Signs: Feeding signs are most obvious, but it creates runways through waterside vegetation. Lives in burrows about 2 cm across; in contrast to rodent burrows the entrance is not worn bare by the animal's feet. **Feeding signs:** Leaves small piles of snail shells, caddisfly cases and partially eaten amphibians or fish on streamside rocks and other prominent places. **Tracks:** Rarely seen. Very small: prints of fore feet 1·2 cm long × 1 cm wide; hind feet 1·4 cm × 1 cm. Like other shrews has five digits: three pointing forwards and two splayed either side; fore and hind prints close together. **Droppings:** Larger than those of Common Shrew, 5–10 mm long and 2–3 mm wide, more oval and with blunter ends. Black when wet, greyish when dry, often with visible fragments of crustaceans. As with other shrews, granular, and flaky when dry – those of rodents are hard.

The strikingly two-toned body and tail is distinctive when the animal is dry …

WHERE TO LOOK/OBSERVATION TIPS

Extremely difficult to observe in the wild, although watching a suitable watercourse (*e.g.* a quiet ditch) on a windless evening might produce results – but at a distance almost impossible to tell from any other swimming small mammal. Can be caught in live-traps (see *page 35*) placed in vegetation close to water (within 2–3 m).

Habitat: Found on the banks of fast-flowing freshwater streams and rivers, among dense vegetation. Also in and around slow-flowing or still water in ponds, canals, marshes and watercress beds; and, in Scotland, on rocky beaches. Sometimes forages in nearby gardens, grassland and woodland, travelling up to 3 km.

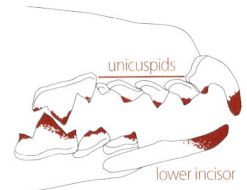

Water Shrew:
four *unicuspid teeth;*
lower incisor blade-like

Food: Feeds both on land and under water, to a depth of 2 m. A voracious predator of aquatic crustaceans and insect larvae, snails, worms, frogs, newts and small fish; and, on land, a range of invertebrates, including beetles, worms and millipedes. Venomous saliva allows it to tackle prey larger than itself.

Habits: Mainly nocturnal, with a peak of activity near dawn. Individuals have a home range along a river bank and adjacent water of typically 60–80 square metres (up to 500 square metres), but are often transient, shifting their home range after a few months. Digs its own burrow system, or takes over old rodent burrows, and may live in close proximity to others of its species, with little aggressive territoriality.

Breeding behaviour: Breeds from April–September, with births peaking in May–June. Each female has one or two litters of 3–15 (average six) young per year. The young are born blind and naked in a nest within the burrow; although they are weaned at 27–28 days, they may remain with their mother for up to 40 days.

Population and status: Population estimated at 1·9 million but believed to be declining due to habitat loss and pollution. Most common in central and southern England. Absent from Ireland.

… but when wet can appear different – the tail looks thinner and the white belly may not be obvious.

See also: Tracks *p. 47*

Lesser White-toothed Shrew

Crocidura suaveolens

(Scilly Shrew)

A relatively pale, reddish-brown shrew found only on the Isles of Scilly in the region covered by this book (although it is absent from some of the smaller islands in the archipelago). It is almost certainly an ancient introduction to Scilly, perhaps by Iron Age traders. (Also occurs on Jersey and Sark in the Channel Islands.)

LEGALLY PROTECTED (UK)	
Introduced (prehistoric); rare (Isles of Scilly only)	
HB:	5·0–7·5 cm
T:	2·4–4·4 cm
Wt:	3–7 g

Identification: As the only shrew on Scilly separation from other species does not pose any identification problems. It has larger, more prominent ears than other British shrews, and its tail is covered with long whitish hairs that stick out at right angles. The teeth are white (most other shrews have a red pigment) and the second unicuspid (single-cusped) tooth is smaller than the third, distinguishing it from the Greater White-toothed Shrew (*page 108*), see *opposite*. However, such details are only evident in the hand, for which a licence is required (see *pages 312–313*).

Sounds: High-pitched squeaks from the undergrowth, and at very close range soft twittering noises may be heard.

As the only shrew on Scilly, readily identified; the large ears are a key feature.

WHERE TO LOOK/OBSERVATION TIPS

Beaches, even if busy, are a good place to look, and dusk and dawn the best times. Sit and wait for animals to scurry into view between rocks and seaweed at the top of the beach.

Signs: Include droppings and tracks, and occasionally food (sandhopper) remains in piles among seaweed. **Droppings:** Very small (0·2–0·4 cm long; 0·1–0·2 cm wide), pointed at one end, and usually black, with a musky smell. **Tracks:** Very small: prints of fore feet 0·8 cm long × 0·9 cm wide; hind feet 1 cm × 1 cm. Prints have five toes (those of small rodents have four).

Habitat: Found wherever there is suitable thick cover, including in gardens. Favoured habitats include heathland and the shoreline, where it forages among boulders and seaweed.

Food: A predator of a range of invertebrates, including beetles, centipedes and millipedes, worms and flies, and sandhoppers and other crustaceans when feeding on beaches.

Habits: Mostly nocturnal, but also active by day, and especially at dusk. Secretive, as with most shrews, and makes tunnels through leaf-litter. Builds a nest of dried grass and twigs, under a log, in an abandoned rodent hole or between boulders. Although individual home ranges are small and overlap, this species is not as aggressively territorial as other shrews.

Breeding behaviour: Breeds March (first mating) to September, producing 2–4 litters of 1–5 (average three) young. The sexes are promiscuous, meeting only to mate. If a litter is disturbed, the mother will lead the young away from the nest in a line, each individual holding on to the base of the tail of the one in front with its teeth. This behaviour, known as 'caravanning', is much more frequent in this species than other British shrews.

Population and status: Estimated between 40,000 and 100,000 individuals, trend unknown but not thought to be threatened.

White-toothed shrews compared

	Lesser White-toothed Shrew	Greater White-toothed Shrew
TEETH:	White. On the upper jaw, three unicuspid (single-cusped) teeth (U1–U3)	
	U2 is distinctly smaller than U3 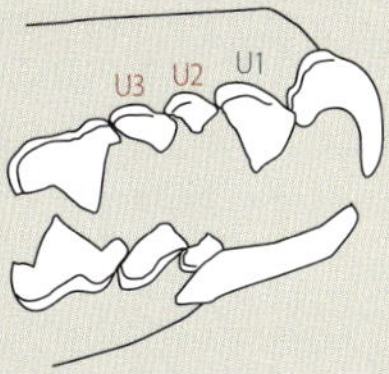	U2 and U3 are approximately the same size. 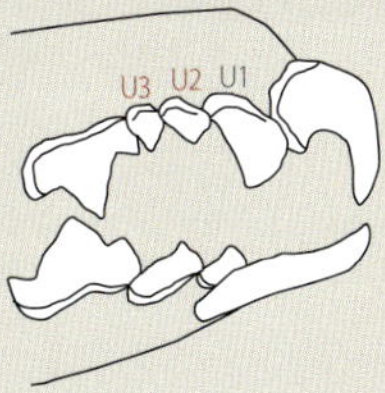
	U3 U2 U1	U3 U2 U1
MANDIBLE LENGTH:	<1 cm	>1 cm
SKULL LENGTH:	<1·85 cm	>1·85 cm

See also: Tracks *p. 47*

Greater White-toothed Shrew

Crocidura russula

(House Shrew)

Although common in mainland Europe, this species was first recorded, in Ireland, in 2007, when skulls were found in Kestrel and Barn Owl pellets. Most likely a recent introduction, it has subsequently been recorded in twenty-seven 10 km squares, and is still spreading. (Also occurs on Guernsey, Alderney and Herm in the Channel Islands.)

Introduced; rare (Ireland only)	
HB:	4·4–8·6 cm
T:	2·4–4·7 cm
Wt:	5–16 g

Identification: The only other shrew in Ireland is the smaller, darker Pygmy Shrew (*page 102*). It is further distinguished from that species by the sparse, erect hairs on its tail, the flatter forehead and more obvious ears, and the lack of red pigment on its teeth. Slightly larger than Lesser White-toothed Shrew (*page 106*), but reliably identifiable only on dental features: its second and third unicuspid (single-cusped) teeth are similar in size, see *page 107*. However, such details are only evident in the hand, for which a licence is required (see *pages 312–313*).

Sounds: Audible high-pitched squeaks from the undergrowth and, at close range, soft twittering sounds.

Signs: Include droppings and tracks, indistinguishable from those of other shrews. Builds an open, saucer-shaped nest of dried grass, under logs, in grass or in compost heaps. **Tracks:** As other shrews.

The larger of the two shrew species found in Ireland.

WHERE TO LOOK/OBSERVATION TIPS

As with all shrews, hard to see as it hides in thick cover. Sitting quietly and waiting in suitable areas within its known range may pay prove successful.

Habitat: Originally discovered in farmland with a mosaic of woodland, grassland and hedgerows, it appears to be particularly frequent around rural settlements.

Food: In common with other shrews, carnivorous, feeding on a wide variety of invertebrates, including woodlice, centipedes, caterpillars, small snails and spiders. Also occasionally takes small vertebrates, including young rodents.

Habits: Active all year and at any time of day and night. Home ranges overlap, and individuals may share nests in winter.

Breeding behaviour: The breeding season in the Channel Islands is from February–October, and is probably the same in Ireland. The sexes are promiscuous and meet only to mate. Females can produce several litters in quick succession, each with 2–10 young. As with Lesser White-toothed Shrew, if a litter is disturbed, the mother will lead the young away from the nest, in a line, with each individual clasping the base of the tail of the one in front with its teeth ('caravanning').

Population and status: Numbers unknown, but apparently common in parts of Co. Tipperary in Ireland, and still spreading. Given the availability of suitable habitat elsewhere in Ireland, it seems very likely that its range will continue to expand. Recent evidence suggests that it is having a negative effect on the population of Pygmy Shrews. Although not specifically protected by domestic legislation, the Greater White-toothed Shrew is covered by some aspects of EU legislation (see *page 314*).

White teeth, lacking red pigmentation, is a key feature of both white-toothed shrew species.

Mole

Talpa europaea

(European Mole)

Molehills and runs are very familiar signs of the presence of Moles, but the animals themselves are rarely seen alive. Inhabiting a tunnel system below ground, adults come to the surface usually only when the ground is too hard or waterlogged. Despite a long history of persecution, owing to the damage they cause to lawns, gardens and playing fields, the Mole is still common, including in woodland where its activities are easily overlooked.

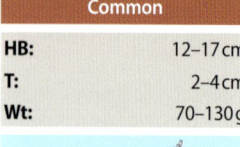

Common	
HB:	12–17 cm
T:	2–4 cm
Wt:	70–130 g

Identification: Supremely adapted to a life of burrowing, the Mole has a cylindrical body, covered in dense (usually black) velvety fur, a long pink snout, minute eyes, and lacks visible ears. The large, spade-like forelimbs, with powerful claws, are rotated such that the inner edge of the foot is turned downwards, and the sole faces backwards. An additional bone gives the impression of a sixth toe.

Sounds: Shrill twitters and squeaks are produced during aggressive encounters with neighbours.

Signs: Molehills, runs and tracks. **Molehills:** Obvious, roughly conical piles of earth of variable size, with no distinct hole on top, often forming lines: they are the spoil heaps from burrowing activity close to the surface. Larger hills (up to 1 m wide and 0·5 m high), known as 'fortresses', often lie above a nest, and may also be built in areas with a high water table. **Runs:** Ridges of earth 4–5 cm high caused by individuals tunnelling just beneath the surface; easy to overlook.
Tracks: Very small but distinctive, although rarely seen. Hind foot print is 1·5 cm long × 1 cm wide, but the fore print is reduced to a series of dots: it walks on the front edges of its front feet so that only the claws leave an impression.

Usually emerge above ground in daylight only during periods of flooding.

WHERE TO LOOK/OBSERVATION TIPS

With much patience it is possible to see the snout of a Mole emerging from an active burrow system. Animals are displaced by floods when they may be seen above ground, and the young disperse overground in July–August, albeit at night.

Habitat: Common and obvious in open areas such as grassland, gardens and agricultural fields, but also in woodland and hedgerows. Tends to avoid sandy and waterlogged soils in which burrowing is difficult.

Food: The underground passageways act as traps for soil invertebrates. Earthworms comprise the bulk of the diet (90% in winter), but almost any invertebrate can be taken, including insects and their larvae, centipedes and millipedes. Moles sometimes disable worms by biting off their heads, and then store them for later consumption (up to 1,200 have been found in a single larder).

Molehills are the most conspicuous sign of the presence of Moles.

Habits: Active year-round in a burrow system with a surface range of 1,300–3,400 square metres, from just below the surface to a depth of 1 m or more, sometimes 'multi-storied'. Highly territorial, an animal will defend its burrow aggressively against intruders. Within each territory is a nest of grass and leaves, to which the animal returns three times a day for regular rests of 3–4 hours.

Breeding behaviour: The breeding season begins in April, with births in May and June. Prior to mating, males extend their tunnels into the territories of neighbouring females, detected underground by smell and sound. There is a single litter of 3–4 young, born naked and blind; their fur starts growing at 14 days. The young disperse at 5–6 weeks, making their way above ground to a new territory.

Population and status: British population estimated at 31 million individuals; although persecuted for centuries, it is not under any threat. Absent from Ireland.

With patience, you may see a Mole stick its head briefly above ground when it is excavating burrows and producing a molehill.

See also: Droppings *p. 42* | Tracks *p. 47*

Hedgehog

Erinaceus europaeus

(Western Hedgehog)

Britain's only spiny mammal, once widespread and familiar almost everywhere, the Hedgehog has suffered a drastic population decline in recent years. It was not recorded in Ireland until the 13th century and may have been introduced there, as it has been to several offshore islands, including the Isles of Scilly, the Uists, Shetland and North Ronaldsay.

LEGALLY PROTECTED	
Biodiversity List (En, Wa, Sc, NI)	
Common	
HB:	30 cm
T:	2 cm
Wt:	450–1,500 g

Identification: Unmistakable, owing to the covering of spines (modified hairs) over the upper side of the body. An adult has about 6,000 spines, usually pale brown, with a darker band towards the tip. The face and underparts are clothed in coarse grey or brown hairs, and the rotund body and very short tail give it a very distinctive shape.

Sounds: Makes a variety of snuffling noises as it moves and forages. Louder grunts and snorts are produced during aggressive encounters and especially when mating.

Signs: Look for nests, tracks and droppings. Broken up horse droppings and cowpats may also indicate foraging activity. **Nests:** Winter nest (hibernaculum) constructed on the ground from leaves, up to 50 cm in diameter and lodged under a log, pile of brushwood, shed floor or in a thicket. Summer day nest similar but smaller; construction is more haphazard and sometimes includes grass and other debris as well as leaves.

The Hedgehog was once a familiar sight, but numbers have declined significantly in recent decades.

WHERE TO LOOK/OBSERVATION TIPS

Suburban areas remain a stronghold, and Hedgehogs can be seen by torchlight especially in gardens where food is provided. Listen for rustling and snuffling noises as they forage. Often seen crossing roads late at night. Footprint tunnels can be used to confirm their presence (see *page 41*).

Tracks: Small: prints of fore feet 2·5–4 cm long × 2·5 cm wide; hind feet 2·5–4·5 cm × 2·5 cm with a stride of 15–25 cm. Prints have five digits, but the inner toe leaves only a weak imprint; the claws are rather long. The fore feet especially are broader and more splayed than a similar-sized Common Rat print (see *page 46*). **Droppings:** 1·5–5 cm long × 1 cm wide, cylindrical. Usually dark grey or black, hard and compressed, and pointed at one end. Shiny insect remains such as beetle wing cases are often visible. Often smell strongly of ammonia. Deposited anywhere, not in a latrine, and frequently found on garden lawns.

Habitat: Requires dense cover for concealment and grassland for foraging, so favoured habitats include woodland edges, scrub and golf courses. Suburban gardens also often provide the right mix of features, but their value can be limited by the presence of fences and walls which restrict foraging movements.

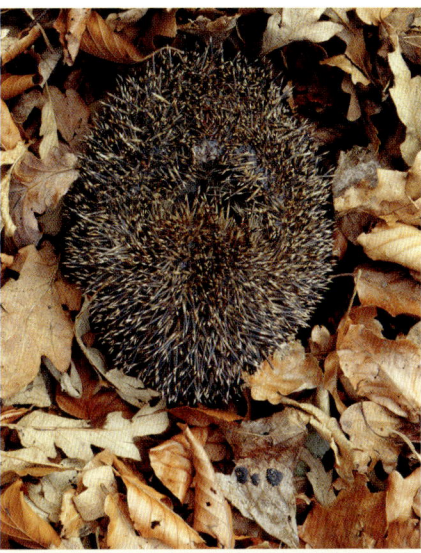

Curls up tightly, with spines raised, when threatened.

Can appear unexpectedly long-legged.

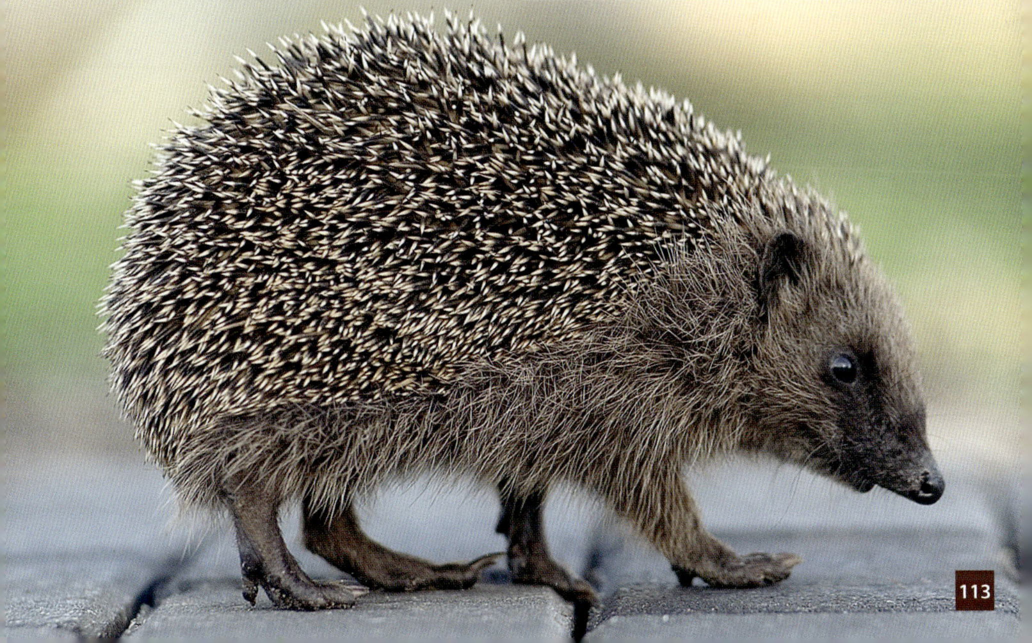

Food: Carnivorous, preying mainly on invertebrates foraged from the ground, with beetles, caterpillars and earthworms especially favoured, but also centipedes, millipedes and slugs. Hedgehogs also feed on the eggs and chicks of ground-nesting birds and carrion, and in autumn will take soft fruits and fungi. Can be attracted to gardens by putting out cat (or dog) food; should not be fed bread and milk as this is indigestible and can lead to the animals' death.

Habits: Mainly nocturnal, but in times of food shortage can be seen during the day. Hibernates from November–March in a nest or sometimes in a Rabbit hole, but may relocate from time to time. Largely solitary, it is not territorial: in a night's foraging in summer it may travel as much as 3 km. In addition to a slow shuffling gait while feeding, the surprisingly long legs enable it to run briskly, climb over obstacles and swim freely. When threatened, it erects its spines and rolls up in a ball: this provides excellent protection against most predators, apart from Badgers, but is one reason why so many are killed by cars on roads.

Breeding behaviour: The breeding season runs from the first matings in April to the last pregnancies in October. Both sexes are promiscuous and the males take no part in caring for the young. The litter of 4–6 young is usually born in June, with another peak in September, although the young from late litters have to hibernate underweight and rarely survive. The young are born within the nest, blind and with white spines visible almost immediately, although these soon turn brown. The young are weaned at 4–6 weeks, leaving the natal home immediately.

Population and status: There are currently estimated to be about 1 million Hedgehogs in Britain, a decline of 50% since the 1990s, and from 36 million in the 1950s. This decline has been attributed to a range of factors including habitat loss and fragmentation, poor hedgerow management, road deaths, flooding, the use of garden strimmers and pesticides (especially molluscicides), and the increase in Badgers, a known predator. The number in Ireland is unknown. Projects are underway to cull or relocate Hedgehogs from some offshore islands in order to protect ground-nesting birds.

Hedgehogs are not at all secretive, and often make snorting and snuffling sounds as they forage.

▲ Hedgehogs hibernate over winter, rolled up in a ball and typically 'burying' themselves …
▼ … in a pile of dead leaves – care should therefore be taken when clearing leaves in the garden.

See also: Tracks *p. 47* | Droppings *p. 42*

Rabbit

Oryctolagus cuniculus

(European Rabbit)

Probably the easiest wild mammal to see in Britain and Ireland, the Rabbit abounds over much of the countryside, in fields, hedgerows and grassland, and is frequently active and visible by day. It originates from Spain and Portugal and was introduced here by the Normans, initially kept in warrens for food and fur. It became widespread from about 1750 but numbers reduced drastically from 1953 following the introduction of myxomatosis. Today the population is again very large, although subject to periodic local fluctuations.

Introduced; very common	
HB:	30–40 cm
T:	5–8 cm
Wt:	1·2–2·0 kg

Identification: The familiar shape, with long ears, long hind feet and short 'cotton' tail easily separates the Rabbit from all other British mammals except Brown (*page 120*) and Mountain (*page 124*) Hares. However, it is noticeably smaller than either of these species, and has much shorter legs. The ears are also shorter, and lack black tips. The fur is usually grey-brown, although black variants are found naturally in some areas, and there are numerous other colour forms in captivity which occasionally escape. Rabbits also tend to feed close to cover, often in groups, whereas Brown Hares are often seen far out in fields, and when running a Rabbit holds its tail upright, showing the white underside.

Although long, a Rabbit's ears are shorter than those of a Brown Hare and do not have black tips.

WHERE TO LOOK/OBSERVATION TIPS
Look along the edges of fields at dusk, especially close to known Rabbit holes. Often seen on the verges of rural roads during car journeys, either by day or caught in the headlights by night.

Sounds: Largely silent, but makes a loud scream when under extreme stress, for example when captured by a predator such as a Stoat (*page 220*). Also stamps its hind feet in alarm, as a communal warning to other rabbits above or below ground.

Signs: Various, including burrows, runs, droppings, latrines, feeding signs and cropped grass. **Burrows:** Entrances are usually only about 10 cm across, but can be up to 50 cm. Although burrows can go down vertically, they are typically on slopes and may honeycomb the ground when several are close together. The entrances are marked by bare excavated soil, especially in sandy areas. **Runs:** Regular paths through low vegetation often radiate out from warrens.

Black variants often occur.

Crouches when threatened.

Paw-scrapes: Shallow holes 5–15 cm long made by both territorial scraping and feeding. Droppings often found nearby. **Feeding signs:** Grazes turf very tightly, usually leaving non-palatable plants such as Ragwort and Nettle. Browses the bottom out of gorse and other bushes to a height of about 30 cm, and gnaws bark from the trunks: the upper incisors have grooves in the centre, which leave a narrow strip of bark in the tooth-marks. **Fur:** Cotton-like tufts of brown, grey or white fur may be left on fences and low bushes. **Droppings:** Spherical pellets, 0·7–1·2 cm in diameter. Black or dark brown when fresh, becoming paler and more fibrous as they dry. Often deposited in large numbers in latrines on prominent features or bare ground. **Tracks:** Fore prints rounded, 3·5 cm long × 2·5 cm wide and hind prints elongate, 6 cm × 2·5 cm; impressions of sharp claws usually visible. When moving slowly, leaves a track with fore prints in pairs with hind prints close behind, but when running one hind print is in front of the other. **Eyeshine:** Red, very low to the ground (see *page 27*).

Habitat: Found in many habitats, including farmland, heathland and moorland, sand dunes, scrub and deciduous woodland, especially on well-drained, light soils. Tends to be seen on the edges of fields, close to the cover of woodland or hedgerows.

Food: Herbivorous, taking a wide variety of plant species, but especially leaves and shoots of grasses. It increases the nutrition its gets from its food by re-ingesting some of its own soft, mucus covered faeces directly from the anus. This usually takes place in the burrow.

Habits: Unlike hares, Rabbits live in a system of burrows known as a warren, and emerge above ground to feed. Although essentially crepuscular and nocturnal, they do venture out by day, especially where undisturbed. Individuals have a very small home range, typically 0·3–0·7 ha, up to a maximum of 2 ha, and rarely move more than 50 m from the warren. Each warren, with many interconnected burrows, is occupied by a stable group of up to five adult males and six females. Separate hierarchies develop between the sexes: dominant males have preferential access to females (and thus sire most young), and dominant females monopolise the best burrows, usually in the centre of the warren. Low-ranking females may resort to nesting in a burrow separate from the rest of the group, and suffer high predation rates. Females are sometimes highly aggressive

Unlike hares, Rabbits show the white underside to their tail when running.

towards youngsters that are not their own. Each social group maintains a separate territory although adjacent groups may share grazing areas. Territorial boundaries are marked by paw-scrapings, scent marking (including urine) and prominent piles of droppings, and males are often seen running along their boundary parallel to a rival. At all times Rabbits remain alert to predators and other threats: their large eyes placed on the side of the head gives almost 360° vision.

Juveniles can be seen any time from about February to September.

Breeding behaviour: A female is capable of producing a litter of 3–7 young every month between January and August, but various factors, both environmental (lack of food) and social (low rank, high population density) usually reduce productivity to about 20 young per year per female. Peak pregnancies occur from April–June. Young are born blind and naked in an underground nest of grass, moss and plucked belly fur. They are suckled only once a day, in the evening, after which the entrance to the nest chamber is sealed. Their eyes open at 10 days, and they are weaned at 21–25 days. Young Rabbits are sexually mature at about four months old, but only those born early in the season can reproduce in the same year.

Population and status: Numbers fluctuate annually (peaking in September/October) and over longer periods (particularly due to diseases such as myxomatosis, which at a local level can have a very significant impact, and Rabbit Viral Haemorrhagic Disease). British population currently estimated at 38 million; the number in Ireland is not known. Rabbit grazing has a considerable economic impact, and influences the ecology and appearance of affected landscapes. Indeed, under the Pests Act 1954, owners and occupiers of land in England are required to control rabbits or stop them causing damage to adjoining crops by erecting rabbit-proof fencing.

In contrast to hares, Rabbits are highly sociable and live in warrens – burrow systems shared by a number of individuals.

Brown Hare

Lepus europaeus

(European Hare)

Brown Hares were introduced to Britain in or before Roman times, probably from the Netherlands, and much more recently to Ireland. They share similar farmland habitats to Rabbits, but live entirely in the open, never resorting to burrows. Brown Hares can reach a speed of 70 km/h (45 mph) when fleeing – and are well known for their courtship, when males chase each other and females repel unwanted advances by 'boxing' at potential suitors.

Identification: Although similar in structure to the Rabbit (*page 116*), adult Brown Hares are much larger, with longer limbs and a more intense russet colour. The head and ears are also larger, the ears being as long as the head, with distinctive black tips. When running, the tail is held down, showing a black dorsal surface. More solitary than Rabbit. Mountain Hare (*page 124*) is slightly smaller and more compact, with relatively shorter ears and an all-white tail.

Sounds: Thumps its front feet in challenge to another hare, and its hind feet to warn of a predator. Otherwise silent unless physically attacked.

LEGALLY PROTECTED (Republic of Ireland) Biodiversity List (En, Wa, Sc)	
Introduced; common	
HB:	48–70 cm
T:	7–13 cm
Wt:	2–5 kg

Black tips to long ears usually evident

Easiest to see on wide open arable fields when recently ploughed or when the sward is short. At rest, hares can often just look like clods of earth – so binoculars or a telescope are invaluable. Driving slowly along farm tracks can be productive – if necessary with the landowner's permission, of course.

Signs: Signs to look out for include forms, trails and droppings. **Forms (nests):** In bare ground, the animal simply scrapes a depression up to 10 cm deep. In tall vegetation, depressions are shallower, consisting mainly of flattened grass. **Trails:** Well-used trails across fields are characterized by 'bound marks' where continuous impacting of the hind feet gives an irregular effect to the path surface. Wears gaps through hedgerows. **Tracks:** Prints of forepaws small, rounded, 4 cm × 4 cm; hind feet prints elongate, 15 cm long × 4·5 cm wide. When hopping slowly, the long hind prints rest side by side behind the fore prints, but when bounding, the tracks are reversed, with hind prints in front of the fore prints. At high speeds the tracks are more separated. **Droppings:** Usually larger than Rabbit droppings, 1·25–2·0 cm in diameter, more flattened and brown rather than black. Surface often rough from plant fibres.
Feeding signs: Gnaws bark from the base of trees or bushes. **Hair:** Often left on fences and low bushes, and might be found in the form during spring and autumn when the animal moults. Moulted fur tends to show more black pigment than that of Rabbits.

Habitat: An animal of open areas; ventures farther out into fields than Rabbits usually do. Usually associated with arable land used for cereal cultivation, especially a patchwork of different crops and fallow ground. Also uses long grass, hedgerows and woodland for resting.

Food: Grazes wild grasses all year (90% of winter diet); also feeds on herbs (50% of summer diet) and some agricultural crops, particularly tender early-growth cereals, winter root crops, peas, beans and maize. Eats its own droppings, taken directly from the anus while resting, to extract more nutrients the second time the material passes through the gut.

Holds ears behind head, sometimes flat along back, when threatened.

Habits: Feeds largely at night, although also active well before sunset and after sunrise, especially in summer. During the day spends most of its time in the form; if flushed it will often run away for a considerable distance (*e.g.* over the brow of a hill). Hares are solitary, but tolerant of others unless their food source is highly localized; each occupies a home range of 20–190 ha and may travel several kilometres in a night. Avoids predators by out-running them, or by sitting tight when the threat is not perceived to be imminent.

Breeding behaviour: The expression 'Mad as a March Hare' arises from the courtship behaviour, which occurs throughout the breeding season, usually March–July. Females are in oestrus for just one day a month, and a hierarchy arises among males competing for receptive females. Dominant males are responsible for most matings, and guard females closely, chasing subordinate males away. The females also try to repel unwelcome males, 'boxing' them with their front legs while standing on their hind legs. Up to three litters are produced in the course of a year, each of 1–4 young (leverets) which are born in a form often lined with the mother's fur. The young are well-furred and have eyes open from birth: they begin to move away from the form after just a few days. The mother leaves the young after giving birth and returns only once a day, at dusk, to suckle. The leverets begin to graze after 13–17 days but are dependent upon their mother's milk for 1–2 months; they first breed in the year after their birth.

Population and status: Current British population estimated at 817,000, fewer than 50 years ago, although the decline is difficult to quantify. Intensification of farming practices may have had an impact, as cereal monocultures provide little food in summer and autumn. Hares also suffer from pesticide pollution and are killed by farm machinery. In Ireland there are about 2,000 animals, and hybridization with Mountain (Irish) Hare has recently been confirmed.

The Brown Hare holds its tail down when running, showing the black upperside (a Rabbit runs with its tail up, showing the white underside).

▲ If you see hares boxing, it is usually a female hitting a male to repel his advances.
▼ Unlike Rabbits, Brown Hares do not dig burrows, instead using shallow depressions in the ground known as forms.

Mountain Hare
Lepus timidus

(Irish Hare)

HB:	45–56 cm
T:	6–7 cm
Wt:	2·5–4·0 kg

The native Mountain Hare is smaller than the Brown Hare and found throughout the Scottish Highlands and across Ireland (subspecies *hibernicus* – the 'Irish Hare'), where the population is genetically very distinct. Populations in the Peak District, as well as those on the Isle of Man and many Scottish islands, originate from introductions. British animals generally occur at higher elevations and in harsher climates than Brown Hare, but the Irish Hare, until the 19th century the only hare on the island of Ireland, is found in lowland habitats as well.

Identification: Slightly smaller than Brown Hare (*page 120*), with a more compact, rounded build and distinctly shorter ears (shorter than the head). At all times differs from Brown Hare by its all-white tail. British animals are grey-brown in summer and the underfur is bluish, which in some light shows clearly, especially on the flanks. Irish Hares are reddish-brown and closer in colour to Brown Hares. In England and Scotland the coat turns white in midwinter (December–February), making it unmistakable; during the moult in October–November and March–May this white is patchy. In Ireland the coat does not go completely white in winter, but always retains brownish patches on the head and back.

Sounds: Produces a distress scream, although rarely heard.

Signs: The main signs are trails through vegetation, forms, droppings and tracks. **Trails:** Makes narrow tracks through Heather and other vegetation, maintained by trampling and grazing. These run up and down hill, whereas sheep and goat trails generally run at an angle to the slope. **Tracks:** Similar to Brown Hare, but in winter the hind feet are heavily furred and may leave very broad tracks. Fore feet prints 3·5 cm long × 3 cm wide; hind feet prints 13 cm × 3 cm. **Forms (nests):** Shallow depressions, often in bare soil, overhung by clumps of old Heather or similar vegetation. In winter, may simply scrape and then lie in snow. **Droppings:** Not distinguishable from Brown Hare droppings. Spherical, 1·25–2 cm in diameter, with a characteristic rough or puckered surface showing the fibrous remains of plant material. Often greenish-yellow, but darker in winter.

Habitat: In Britain, montane grassland and open heather moorland above 300 m is the main habitat, but also occurs in woodland with a rich ground flora. In Ireland, common in a wider range of upland and lowland habitats.

Food: Heather (especially in winter) and various grasses, sedges and rushes are the mainstays of the diet in Scotland and in the Peak District, whereas grass dominates the diet throughout the year in Ireland.

WHERE TO LOOK/OBSERVATION TIPS

Easiest to see in early spring when their white fur colour stands out against a dark moorland background. Often flushed from a form during hikes in the Scottish mountains. Seen out in the open at dusk and dawn, and also sometimes in bad weather.

▲ In summer, the coat is a variable mixture of grey-brown and white
▼ In winter the coat is wholly white in Britain, but mottled in Ireland

Habits: Mainly nocturnal, but also active at dawn and dusk. Rests in a form on high slopes by day, moving at dusk to night-time grazing sites lower down. Found at higher altitudes in summer than in winter. Each individual has a home range (largest in males), but they are not territorial and often feed together, sometimes in large groups.

Breeding behaviour: A long breeding season begins early, with mating from January and young (leverets) born between March and October. There are up to three litters a year, with 1–3 young in each, born fully furred and with their eyes open. The female suckles the young once a day, at sunset. Leverets start to leave the breeding form after only a week, and are weaned at three weeks. Has similar courtship behaviour to Brown Hare, with males chasing females, which may 'box' to rebuff their advances.

Population and status: British population estimated at 360,000, mostly in Scotland but up to 10,000 in the Peak District. Common in Ireland, although shows considerable natural population fluctuations over a period of 5–15 years: population estimates for the Republic of Ireland of 233,000 in 2006 and 535,000 in 2007, and for Northern Ireland of 59,700–86,900. Numbers are controlled on some shooting estates in the belief that this inhibits the spread of a viral disease known as louping ill that can be fatal in Red Grouse.

Irish Hare subspecies hibernicus *in typically brown summer coat. Note the much shorter ears and more compact body shape compared with Brown Hare.*

▲ Some individuals are largely grey-brown in summer, but always have white legs.
▼ In Britain, occurs mainly on moorland above 300 m, but widespread in Ireland.

Identifying bats

Bats are remarkable and unique mammals which set particular identification challenges. They are the only mammals that fly and generally do so at night when it is difficult or impossible to see them, let alone identify them to species. During the day and in winter they seek out dark, sheltered places to roost, where they can also be hard to find.

Bats are vulnerable mammals, relying on catching flying insects in fickle weather and finding safe, undisturbed roost sites to survive and reproduce. Many species in Britain and Ireland declined in numbers very significantly during the 20th century and, as a result, all are now strictly protected – which means that, even if you do find some at roost, it is illegal to disturb them.

Fortunately, there are ways to study and appreciate bats without breaking the law. All the species in Britain and Ireland echolocate, producing sounds to hunt and navigate. This means that they can forage and commute in the dark, avoiding being hunted by diurnal birds of prey (although they are taken by owls), and it also means that we can detect them as they do so. Anybody can go out at night and listen to calls with a bat detector (see *pages 134–137*), and with a little practice most people can learn the sounds of the commoner species.

When bats are not hunting, they return to 'base', their roosting site. They roost in trees and caves, and also in buildings and other structures, often living alongside people. To visit or monitor a roosting site you must do so in the company of a licensed bat worker, both to ensure the welfare of bats and to avoid breaking the law. Different species have different types of roost, and the same individuals can also inhabit a number of roosts throughout the year, with maternity roosts and hibernation roosts being particularly important sites. Licensed bat workers regularly undertake visits to roosts, which can involve handling the bats, and with their agreement people without a licence can also be present to observe. On the following pages is a quick guide to identifying bats in the hand, more detailed information on identification being provided in the species accounts.

Of course, you might find that there are bats in your own house, or in buildings nearby. If you know the location of a roost, bats can generally be watched emerging from or re-entering the roost – although to minimise disturbance you should do so from a distance of at least five metres and avoid the use of bright lights. The entrance and exit points for roosts can be small and the signs of bats can be subtle, comprising droppings, scratches at the entrance, and sometimes some staining. The size and shape of bat droppings can provide a useful clue to the species present, and a guide to identifying bats by this means is included on pages *138–139*.

There is an active network of local bat groups across Britain and Ireland that encourage people with an interest to get involved, whatever their level of knowledge. The Bat Conservation Trust (see *page 319*) runs the National Bat Monitoring Programme, which is a citizen science project that includes activities for the beginner, such as simply recording the bats present in a garden or park, as well as more involved activities such as transect surveys. The 'Bat Helpline' (+44 (0)345 1300 228) is also an excellent source of advice, especially for people who have bats roosting in their house or outbuildings.

All bats are legally protected in Britain and Ireland and under the relevant domestic legislation (see *pages 312–313*) a licence is required to disturb them at their roost, or to handle them. Obtaining a licence requires training, although the process involved depends on the country concerned. For information, contact the relevant Statutory Nature Conservation Body (SNCB) (see *page 318*).

Naming the parts

The secret to learning how to identify any group of species is first to obtain some knowledge of the features that are used to separate them. In the case of bats, the key features are often subtle and only possible to see when the animal is examined in the hand (which requires a licence). The technical terms used in this book are shown on the annotated photographs below.

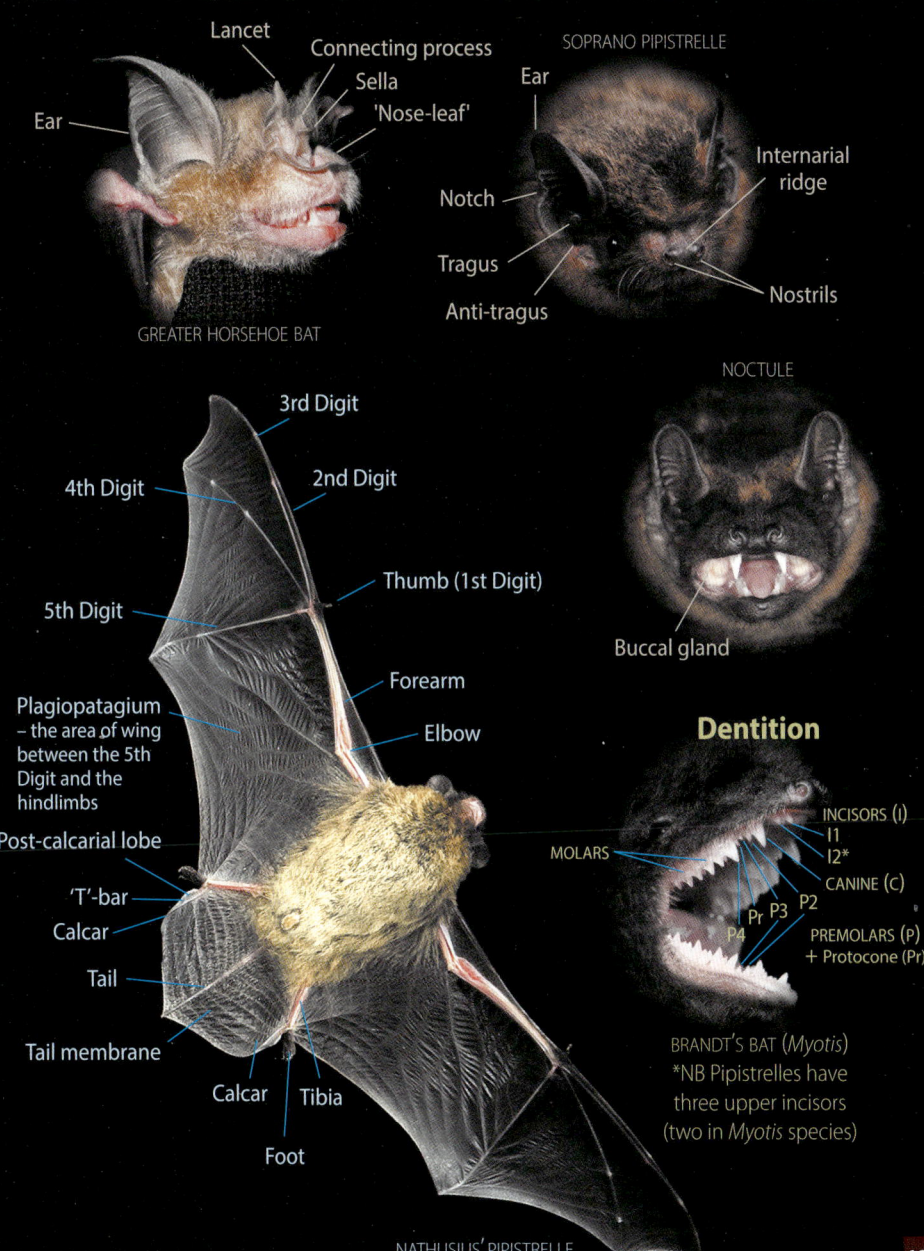

Lancet
Connecting process
Sella
'Nose-leaf'
Ear

GREATER HORSESHOE BAT

SOPRANO PIPISTRELLE

Ear
Internarial ridge
Notch
Tragus
Anti-tragus
Nostrils

3rd Digit
2nd Digit
4th Digit
Thumb (1st Digit)
5th Digit
Plagiopatagium – the area of wing between the 5th Digit and the hindlimbs
Forearm
Elbow
Post-calcarial lobe
'T'-bar
Calcar
Tail
Tail membrane
Calcar
Tibia
Foot

NATHUSIUS' PIPISTRELLE

NOCTULE

Buccal gland

Dentition

INCISORS (I)
I1
I2*
CANINE (C)
MOLARS
Pr P3 P2
P4
PREMOLARS (P) + Protocone (Pr)

BRANDT'S BAT (*Myotis*)
*NB Pipistrelles have three upper incisors (two in *Myotis* species)

129

Key to the breeding bats of Britain and Ireland

The following is a simple, step-by-step guide to the identification of the 17 bat species breeding in Britain and/or Ireland.

To confirm identification a bat may need to be examined in the hand in order to assess and measure key features. It is important to note that a licence is required to handle a bat and as such can only be done by a suitably qualified person.

Summary descriptions of the nine species that have been recorded as rare migrants or vagrants are included on *page 133*: those that may be confused with the breeding species are highlighted in this identification key using a 'Rare Beware' symbol ◉ .

Bats with a horseshoe-shaped 'nose-leaf'

Large bat – size of a pear
Greater Horseshoe Bat (*page 140*)

Small bat – size of a plum
Lesser Horseshoe Bat (*page 144*)

Horseshoe bats

Use nose shape for initial differentiation

Ears widely spaced – do not meet in middle of forehead

Bats without a 'nose-leaf'

Ears meet (or almost meet) in middle of forehead

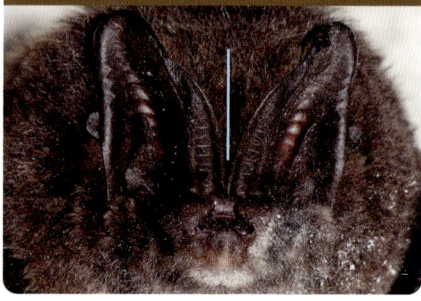

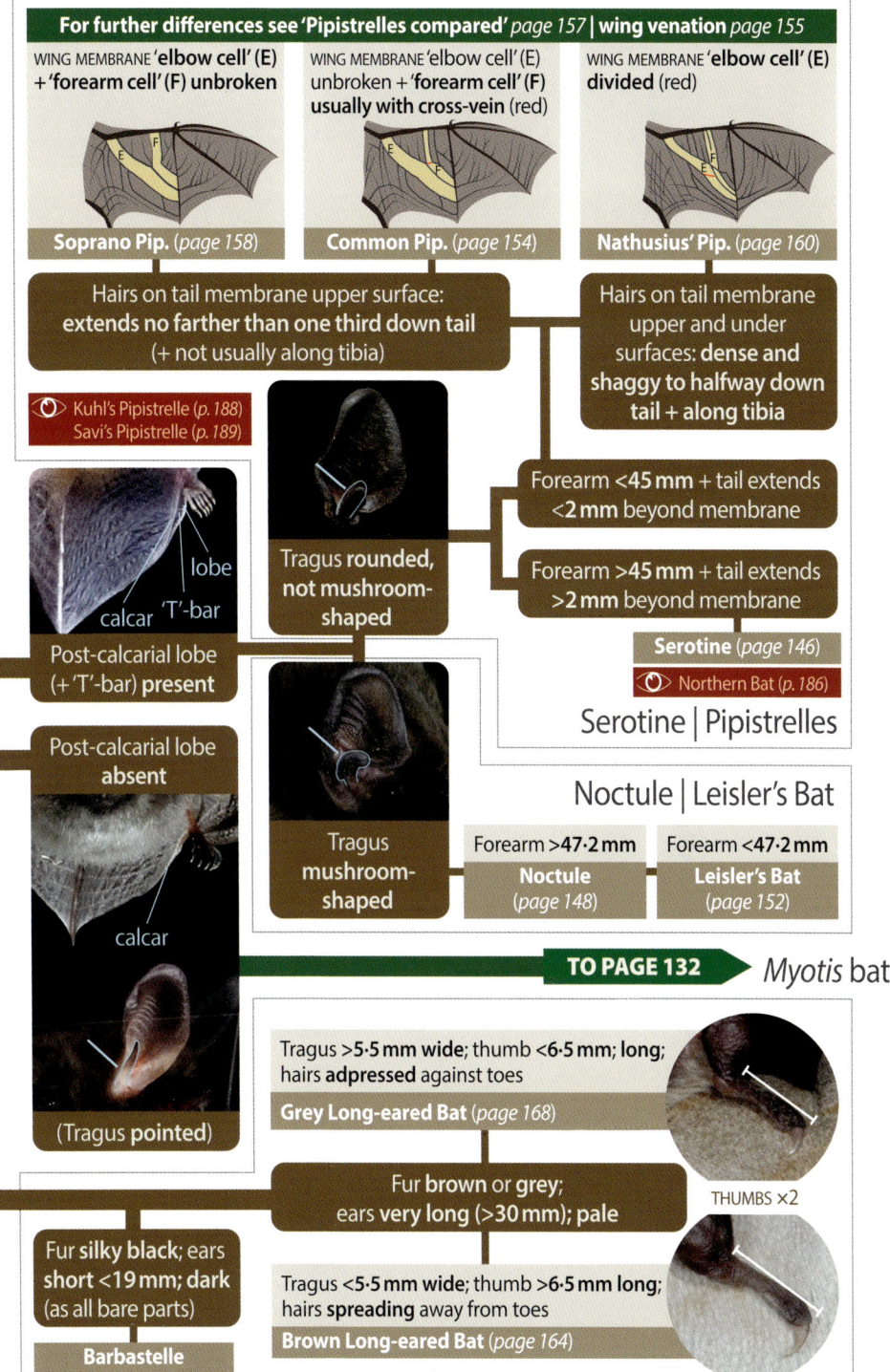

For further differences see 'Pipistrelles compared' *page 157* | **wing venation** *page 155*

WING MEMBRANE **'elbow cell' (E) + 'forearm cell' (F) unbroken**

Soprano Pip. (*page 158*)

WING MEMBRANE **'elbow cell' (E) unbroken + 'forearm cell' (F) usually with cross-vein** (red)

Common Pip. (*page 154*)

WING MEMBRANE **'elbow cell' (E) divided** (red)

Nathusius' Pip. (*page 160*)

Hairs on tail membrane upper surface: **extends no farther than one third down tail** (+ not usually along tibia)

Kuhl's Pipistrelle (*p. 188*)
Savi's Pipistrelle (*p. 189*)

Hairs on tail membrane upper and under surfaces: **dense and shaggy to halfway down tail + along tibia**

Forearm **<45 mm** + tail extends **<2 mm** beyond membrane

Forearm **>45 mm** + tail extends **>2 mm** beyond membrane

Serotine (*page 146*)

Northern Bat (*p. 186*)

Serotine | Pipistrelles

lobe
'T'-bar
calcar

Tragus **rounded, not mushroom-shaped**

Post-calcarial lobe (+ 'T'-bar) **present**

Post-calcarial lobe **absent**

Tragus **mushroom-shaped**

Noctule | Leisler's Bat

Forearm **>47·2 mm**	Forearm **<47·2 mm**
Noctule (*page 148*)	**Leisler's Bat** (*page 152*)

calcar

TO PAGE 132 ➤ *Myotis* bats

(Tragus **pointed**)

Tragus **>5·5 mm** wide; thumb **<6·5 mm**; long; hairs **adpressed** against toes

Grey Long-eared Bat (*page 168*)

Fur **brown** or **grey**; ears **very long** (>30 mm); pale

THUMBS ×2

Fur **silky black**; ears **short <19 mm**; **dark** (as all bare parts)

Barbastelle (*page 162*)

Tragus **<5·5 mm** wide; thumb **>6·5 mm** long; hairs **spreading** away from toes

Brown Long-eared Bat (*page 164*)

Barbastelle | Long-eared bats

FROM PAGE 131

Bats without 'nose-leaf' | ears widely spaced | post calcarial lobe absent + tragus pointed

Myotis bats

Ears **>18 mm** long (when pushed forward, they extend well beyond nose tip)

Bechstein's Bat (*page 170*)

👁 Greater Mouse-eared Bat (*p. 184*)

Ears **<18 mm** long (when pushed forward, they usually extend no farther than nose tip)

Calcar (c) usually straight; **short** – <50% of distance from foot to tail (ft)

Calcar (c) usually straight; **long** – approx. 75% of distance from foot to tail (ft); soft hairs along edge of tail membrane

Calcar **'S'-shaped**; stiff bristles along edge of tail membrane

Natterer's Bat (*page 172*)

👁 Geoffroy's Bat (*p. 192*)

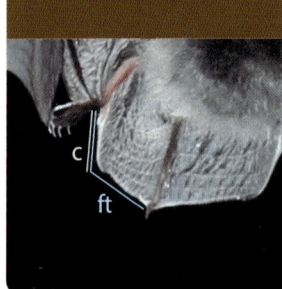

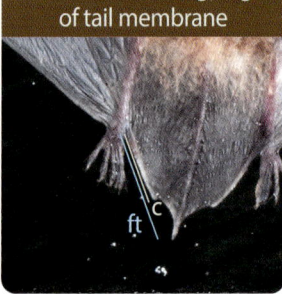

Daubenton's Bat (*page 174*)

👁 Pond Bat (*p. 193*)

For further differences see 'the 'WAB' group' compared' *page 177*				
	Dentition	Nostrils	Tragus	Penis
Whiskered Bat (*page 178*)	3rd premolar on lower jaw (P3) less than half the size of 2nd (P2); protocone (Pr) very **small or absent**	rounded/ comma-shaped	≥ notch	parallel-sided
Alcathoe Bat (*page 180*)	3rd premolar on lower jaw (P3) less than half the size of 2nd (P2); protocone (Pr) pronounced	heart-shaped	< notch	parallel-sided
Brandt's Bat (*page 182*)	3rd premolar on lower jaw (P3) **slightly smaller** than 2nd (P2); protocone (Pr) pronounced	heart-shaped	≥ notch	swollen tip

Identification of the rare bats of Britain and Ireland

In addition to the 17 species of bat that breed in Britain and/or Ireland, nine have occurred as rare migrants or vagrants. A summary of the identification features is provided in the table below, but in most cases certain identification is only possible by examining the animal in the hand, which requires a licence. Regularly occurring species with which these rarities may be confused are **highlighted**.

Greater Mouse-eared Bat *p. 184*	Readily identified by large size, pale body (light brown above, whitish below) and broad ears with a pointed tragus. If a bat in the hand keys out as **Bechstein's Bat** (*page 170*) but is large, check forearm length (57–71 mm in Greater Mouse-eared Bat; 38–45 mm in Bechstein's Bat).
Northern Bat *p. 186*	Medium-sized: similar to **Serotine** (*page 146*) with black ears and face, but slightly smaller and has rounder ears and darker fur on the upperparts with golden tips, more sharply contrasting with yellowish-brown fur on underparts. Superficially similar to the rare Particoloured Bat (*page 187*) but does not have silvery tips to the fur and lacks white on the underside.
Hoary Bat *p. 191*	Large and very distinctive: thick, 'frosty'-tipped fur gives grizzled appearance, and has an obvious yellowish collar. The rare Particoloured Bat (*page 187*) is similar, but smaller, lacks the yellowish collar and is usually white on the underside.
Particoloured Bat *p. 187*	Medium-sized: 'frosty' tips to long, dense fur on upperparts giving a distinctive silvery appearance. Usually white on the underside and often has a yellowish patch of skin around the ears, which are short and broad; their lower edge extends in a fold below the line of the mouth. The dark face and membranes are similar to those of **Serotine** (*page 146*) and the rare Northern Bat (*page 186*) but these species do not have a silvery back.
Kuhl's Pipistrelle *p. 188*	Slightly larger than **Common Pipistrelle** (*page 154*) and usually paler brown above and creamy-white below. Has paler, reddish-brown ears and face (young are darker) and a sharply defined white trailing edge to the wing, between the foot and the fifth finger. Uniquely among pipistrelles, has only a single cusp on the first incisor.
Savi's Pipistrelle *p. 189*	Very small, like *Pipistrellus* species (*pages 154–161*): differs in dentition; has slightly longer, 'fluffier' fur with distinct contrast between dark upperside and paler underside; wider, rounder ears; blackish and shiny face, ears and wings. Fluffy appearance and colour of fur similar to the larger, rare, Particoloured Bat (*page 187*) and Northern Bat (*page 186*).
Pond Bat *p. 193*	Like a big **Daubenton's Bat** (*page 174*) but has greyer-brown fur, lacking any reddish tinge, more sharply demarcated whiter fur on the underparts and a reddish-brown face (darker in juveniles).
Geoffroy's Bat *p. 192*	Medium-sized, with a very distinctive right-angled fold or notch on the outer edge of the ears. Otherwise similar to **Natterer's Bat** (*page 172*) although lacks the sharp contrast between the upperparts and underparts.
European Free-tailed Bat *p. 190*	Very large, with forward-facing, egg-shaped ears that meet on the head, and relatively large eyes. Fur is greyish-brown but lighter on the underside. Unique (in Europe) in that the tail projects well beyond the membrane.

IDENTIFYING BATS BY SOUND

Many bats can be identified to species using a bat detector, which converts their ultrasonic calls into audible signals or visible graphs of pitch against time (sonograms). There are many types of bat detector available, but the least expensive are 'heterodyne' detectors (see *page 36*). The chart on *pages 136–137* summarises the general characteristics of the 'calls' of each species as heard using a heterodyne detector, with a broad indication of the frequency at which they can be recorded. The frequency of the most powerful (loudest) part of the call is known as the peak frequency (PF) and can be important for identification.

It is important to remember that when using a bat detector it is often not possible to identify species with confidence, especially for bats with overlapping call characteristics (*e.g. Myotis* species). Furthermore, the calls of some species are highly variable depending on the nature of the habitat they are in. As a result, the calls of different species can sound quite similar, and equally may differ significantly from the typical patterns described in this book.

The sonograms here show typical bat calls as recorded on a full spectrum bat detector for some species with distinctly different calls; these calls fall into three broad categories.

Frequency-modulated (FM) calls
This type of call is a sound burst that sweeps down the frequency range and usually starts and ends at characteristic frequencies (pitches) depending on the species, often appearing as a near-vertical line on a sonogram. This type of call is generally heard as clicks on a bat detector.

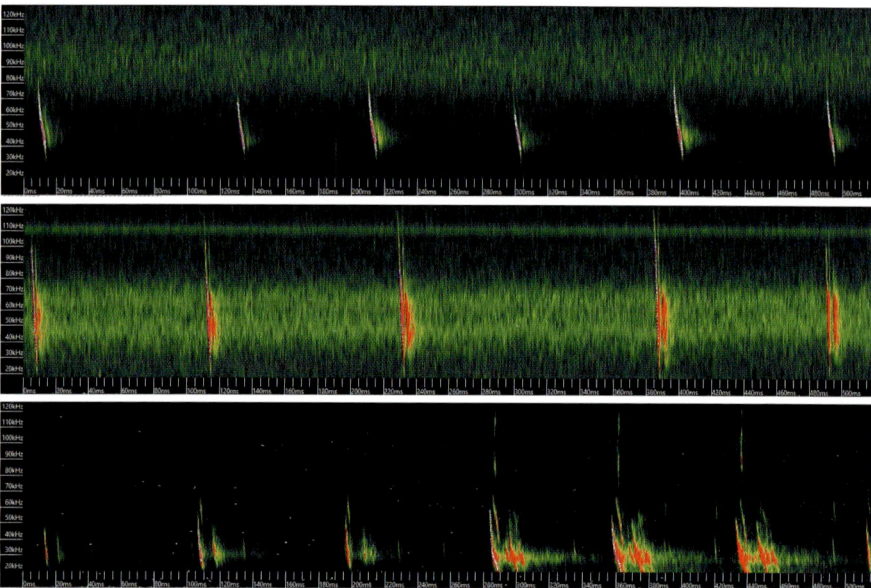

Myotis species, such as Daubenton's Bat (TOP), Natterer's Bat (MIDDLE) and long-eared bats (BOTTOM – Grey Long-eared Bat) have frequency-modulated (FM) calls that are short in duration, cover a broad range and have a short time interval between calls.

Constant-frequency (CF) calls

This type of call is more of a pulse in which the same pitch is maintained throughout. They often last longer than FM bursts, and are particularly distinctive in horseshoe bats.

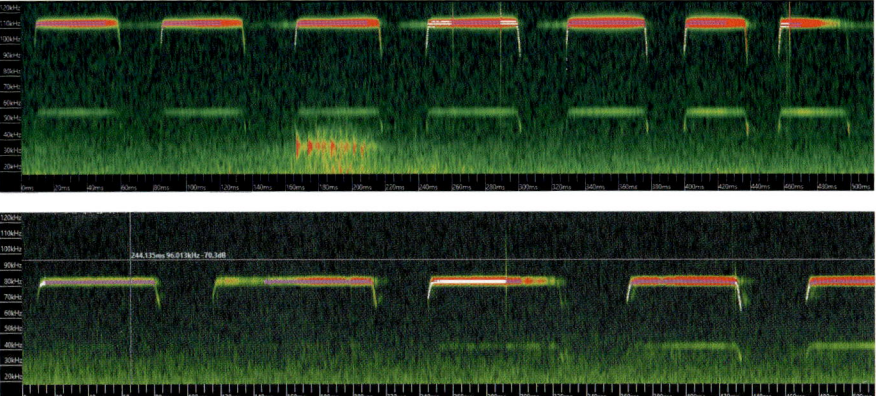

Horseshoe bats (Lesser (TOP); Greater (BOTTOM)) have a very distinctive constant frequency (CF) call.

Quasi constant frequency (QCF) calls (mixed FM/CF calls)

Many calls have a FM burst followed by a CF component, the latter showing up as a 'tail' on a sonogram, sometimes described as 'hockey-stick' shaped. These calls are of longer duration, and on a heterodyne detector the sound can be described as more of a 'smack' than a 'click'. Loudness varies between species, with the long-eared bats making very quiet calls and others, such as Noctule, making very loud calls.

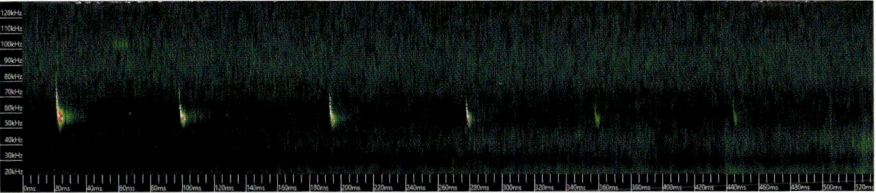

Pipistrelle calls have a 'broadband' frequency-modulated (FM) element, with a constant frequency (CF) tail at the peak frequency.

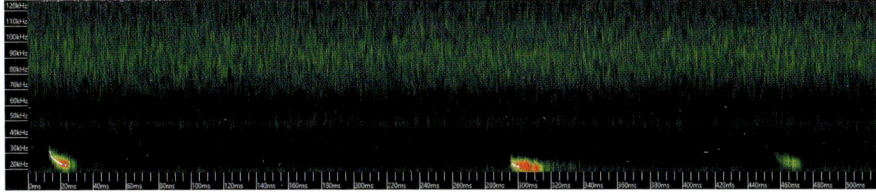

Large bats such as Noctule and Serotine tend to have longer duration, lower frequency calls, with longer gaps between calls. This is broadly related to the larger prey that they eat.

Summary of bat sounds

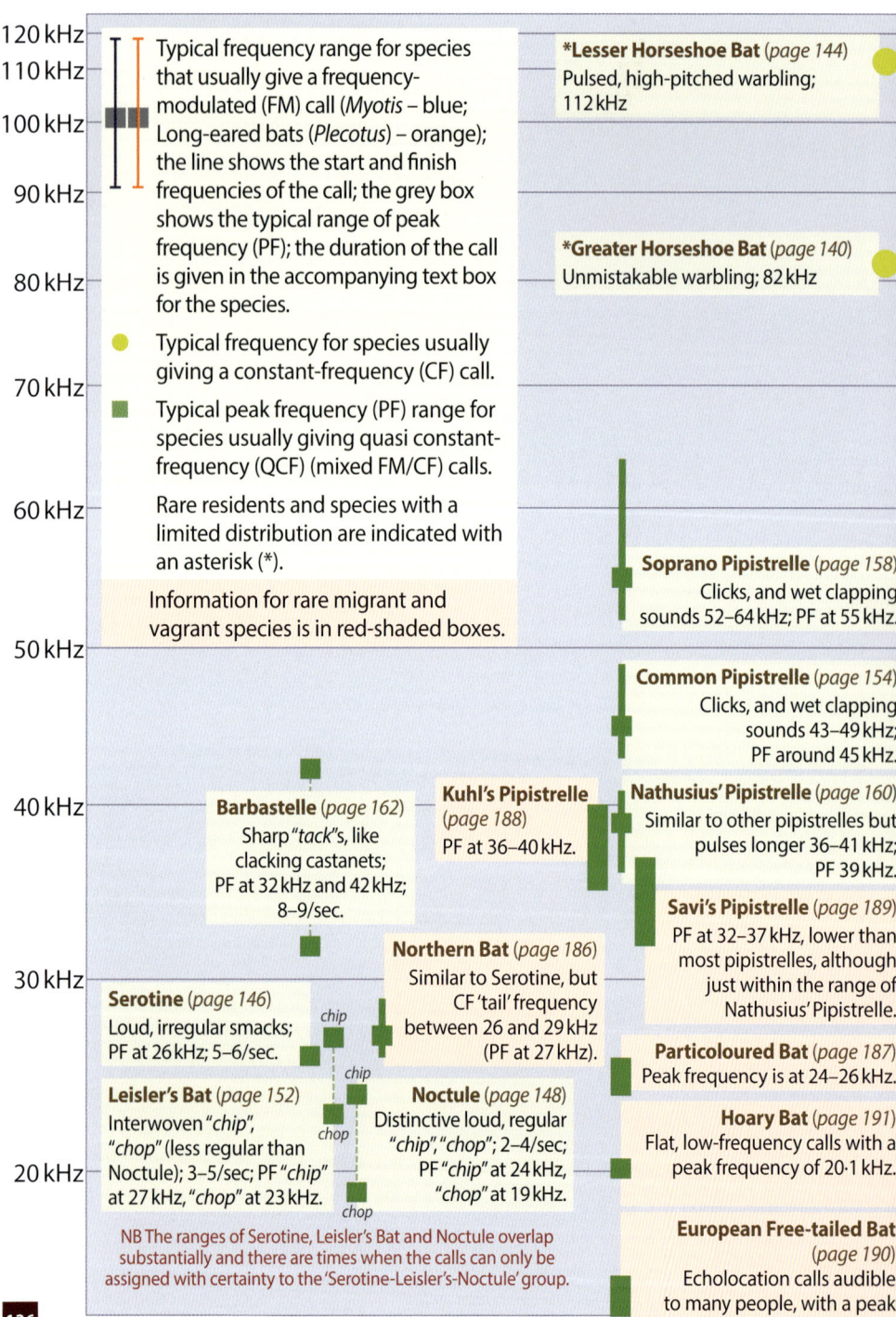

Typical frequency range for species that usually give a frequency-modulated (FM) call (*Myotis* – blue; Long-eared bats (*Plecotus*) – orange; the line shows the start and finish frequencies of the call; the grey box shows the typical range of peak frequency (PF); the duration of the call is given in the accompanying text box for the species.

Typical frequency for species usually giving a constant-frequency (CF) call.

Typical peak frequency (PF) range for species usually giving quasi constant-frequency (QCF) (mixed FM/CF) calls.

Rare residents and species with a limited distribution are indicated with an asterisk (*).

Information for rare migrant and vagrant species is in red-shaded boxes.

*Lesser Horseshoe Bat (*page 144*)
Pulsed, high-pitched warbling; 112 kHz

*Greater Horseshoe Bat (*page 140*)
Unmistakable warbling; 82 kHz

Soprano Pipistrelle (*page 158*)
Clicks, and wet clapping sounds 52–64 kHz; PF at 55 kHz.

Common Pipistrelle (*page 154*)
Clicks, and wet clapping sounds 43–49 kHz; PF around 45 kHz.

Barbastelle (*page 162*)
Sharp "*tack*"s, like clacking castanets; PF at 32 kHz and 42 kHz; 8–9/sec.

Kuhl's Pipistrelle (*page 188*)
PF at 36–40 kHz.

Nathusius' Pipistrelle (*page 160*)
Similar to other pipistrelles but pulses longer 36–41 kHz; PF 39 kHz.

Savi's Pipistrelle (*page 189*)
PF at 32–37 kHz, lower than most pipistrelles, although just within the range of Nathusius' Pipistrelle.

Northern Bat (*page 186*)
Similar to Serotine, but CF 'tail' frequency between 26 and 29 kHz (PF at 27 kHz).

Serotine (*page 146*)
Loud, irregular smacks; PF at 26 kHz; 5–6/sec.

Leisler's Bat (*page 152*)
Interwoven "*chip*", "*chop*" (less regular than Noctule); 3–5/sec; PF "*chip*" at 27 kHz, "*chop*" at 23 kHz.

Noctule (*page 148*)
Distinctive loud, regular "*chip*", "*chop*"; 2–4/sec; PF "*chip*" at 24 kHz, "*chop*" at 19 kHz.

Particoloured Bat (*page 187*)
Peak frequency is at 24–26 kHz.

Hoary Bat (*page 191*)
Flat, low-frequency calls with a peak frequency of 20·1 kHz.

NB The ranges of Serotine, Leisler's Bat and Noctule overlap substantially and there are times when the calls can only be assigned with certainty to the 'Serotine-Leisler's-Noctule' group.

European Free-tailed Bat (*page 190*)
Echolocation calls audible to many people, with a peak frequency at 10–14 kHz.

chip
chip
chop
chop

*Grey Long-eared Bat   GLE
(page 168)

As Brown Long-eared Bat: rapid staccato clicks at 17–56 kHz; PF at 33 kHz.

Brown Long-eared Bat   BLE
(page 164)

Quiet, rapid staccato clicks at 25–70 kHz; PF at 33 kHz.

Greater Mouse-eared Bat   GMO
(page 184)

FM calls fall from about 80 kHz to 28 kHz in about 4·6 ms; PF at 36 kHz.

Daubenton's Bat (page 174)   DAU

'Dry' and fast clicks, like a distant machine gun; falls from 81 kHz to 29 kHz in about 3·2 ms; PF at 47 kHz.

Pond Bat (page 193)   PON

FM calls fall from 85–65 kHz to 35–25 kHz in 4–8 ms; PF at 35–45 kHz. Not safely distinguishable from Daubenton's Bat.

Whiskered Bat (page 178)   WHI

Indistinguishable from Brandt's Bat. Rapid 'dry' clicks; falls from 88 kHz to 32 kHz in about 4·2 ms; PF at about 48 kHz.

Brandt's Bat (page 182)   BRA

Indistinguishable from Whiskered Bat. Rapid 'dry' clicks; falls from 92 kHz to 34 kHz in about 3·5 ms; PF at about 47 kHz.

Natterer's Bat (page 172)   NAT

Short, irregular crackling pulses; falls from 107 kHz down to 23 kHz in about 4·7 ms; PF at 47 kHz.

*Bechstein's Bat (page 170)   BEC

'Dry' clicks; falls from 116 kHz to 33 kHz in about 2·4 ms; PF at 51 kHz. Typically, a relatively quiet call.

*Alcathoe Bat (page 180)   ALC

Falls from 120 kHz to 43 kHz in <4 ms; PF at 53 kHz.

Geoffroy's Bat (page 192)   GEO

FM calls fall from over 140 kHz to about 38 kHz in 1·5–4·0 ms; PF at 55–68 kHz.

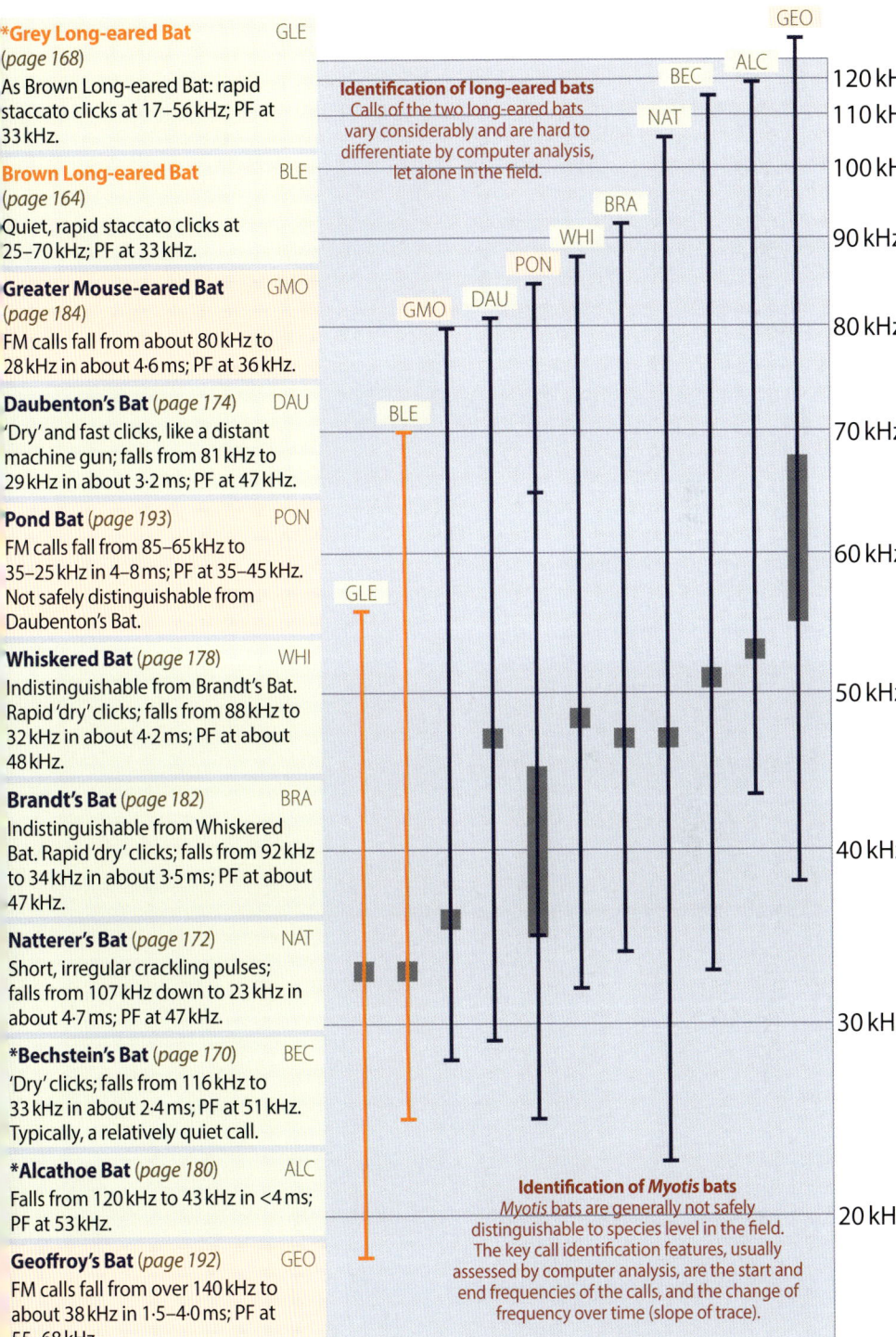

Identification of long-eared bats
Calls of the two long-eared bats vary considerably and are hard to differentiate by computer analysis, let alone in the field.

Identification of Myotis bats
Myotis bats are generally not safely distinguishable to species level in the field. The key call identification features, usually assessed by computer analysis, are the start and end frequencies of the calls, and the change of frequency over time (slope of trace).

GEO · BEC · ALC · NAT · BRA · WHI · PON · GMO · DAU · BLE · GLE

120 kHz · 110 kHz · 100 kHz · 90 kHz · 80 kHz · 70 kHz · 60 kHz · 50 kHz · 40 kHz · 30 kHz · 20 kHz

IDENTIFYING BATS BY DROPPINGS

Bat droppings rarely smell and are generally harmless – and can often be used to identify the species (or type of species) present in a roost. The table *opposite* provides summary descriptions and illustrations of typical droppings for all regularly occurring British and Irish bats. The illustrations are shown at actual size (enabling direct comparison with any droppings you find), and arranged in increasing order of size with the smallest droppings first. It is worth bearing in mind that the droppings of any particular species can vary in size, shape and texture, and ideally a selection of droppings should therefore be collected in order to get a feel for what an 'average' pellet looks like.

The presence of Common Pipistrelles can often be detected by droppings on leaves or on the ground below the entrance to their roost site.

A linear mass of Brown Long-eared Bat droppings concentrated beneath the central ridge board of a roof.

Well-used roosting holes, such as this Noctule roost, may show a dark stain of urine and droppings, and often smell of ammonia.

Species	Particle size	Width (mm)	Length (mm)	Description	Shape (life-size)
Lesser Horseshoe Bat (page 144)	Fine to medium	1·5–2·0	6–8	Often 3–4 ovoids; can be squidgy and contain cranefly legs	
Whiskered, Brandt's & Alcathoe Bats (pages 178–182)	Medium	2·0–2·3	6–9	Like those of pipistrelles but rounder and fatter	
Leisler's Bat (page 152)	Medium	2·5–3·0	6–9	Usually brown, and wetter than those of Noctule	
Pipistrelles (all species) (pages 154–160)	Fine	1·5–2·0	7–9	Smooth, knobbly outline; often pointed both ends	
Daubenton's Bat (page 174)	Fine	1·8–2·3	8–9	Smooth outline; squidgy if fresh; smells of mud	
Barbastelle (page 162)	Medium	2·1–2·7	8–11	Smooth and knobbly outline; in 3 parts	
Natterer's Bat (page 172)	Medium	2·3–3·3	8–11	3-lobed, usually with helical structure	
Long-eared bats (pages 164–168)	Medium to coarse	2·5–3·0	8–11	3-lobed, with knobbly outline	
Serotine (page 146)	Coarse	3·5–4·0	8–11	Ovoid with a 'tail'	
Bechstein's Bat (page 170)	Medium to coarse	2·5–3·5	9–12	Always contains shiny fragments	
Greater Horseshoe Bat (page 140)	Coarse	2·2–2·7	9–13	Usually has shiny fragments; generally comprises 3 ovoids	
Noctule (page 148)	Medium	2·0–3·5	11–15	Usually brown	
Greater Mouse-eared Bat (page 184)	Coarse	3·5–4·5	12–17	Large, dark brown; fragments of beetles etc.	

Horseshoe bats

2 British/Irish species

The two horseshoe bats differ from all other British and Irish bats in a number of ways: they have a horseshoe-shaped flap of skin on their face, which, with the additional structures known as the sella and the lancet (see *opposite*) and the connecting process, make up the 'nose-leaf'; they do not have a tragus; and they echolocate through their nose while the mouth is shut (other British species echolocate with the mouth open). Horseshoe bats have long legs and when at rest always hang free. When torpid, for example during hibernation, they wrap their wings tightly around their body. They are also unique in having 'false nipples' which the young cling on to when at rest on their mother. The two species are readily told apart by their size.

Greater Horseshoe Bat   *Rhinolophus ferrumequinum*

Greater Horseshoe Bat is now very rare in Britain, having suffered a catastrophic long-term decline.

Identification: Second only to Noctule (*page 148*) in size, the Greater Horseshoe Bat is instantly recognisable in the hand and at roost, its face dominated by a pink 'nose-leaf' with horseshoe-shaped folds. Lesser Horseshoe Bat (*page 144*) has similar facial features, but is very considerably smaller, and unlikely to be confused.

Flight: Often flies low (0·3–6·0 m); a distinctive slow, fluttering flight with short glides. Sometimes flies very close to vegetation and walls, its broad wings giving it considerable manoeuvrability.

Sounds: The constant-frequency, very directional calls made by horseshoe bats are translated by a heterodyne bat detector into pulses of distinctive warbling. The peak frequency for a Greater Horseshoe bat is 82 kHz, although harmonics may also be heard around 40 kHz.

Signs: Summer roosts: In roof spaces of old buildings.
Winter roosts: In caves and mines which are not too cold: requires a temperature of over 7°C.

Habitat: Warm areas with a diversity of habitats, the essentials being broad-leaved woodland, pasture (especially with cows) and hedgerows.

LEGALLY PROTECTED	
Biodiversity List (En, Wa)	
Rare	
HB:	5·4–7·1 cm
WS:	35–40 cm
Wt:	17–34 g

Food: Specializes on large beetles (especially dung beetles) and moths. In spring, Common Cockchafer is a major food item, while later in the season big moths, such as Large Yellow Underwings, are preferred.

Habits: Hibernates from October to April, but sometimes wakes up to feed on warm winter nights. In summer, emerges about 30 minutes after sunset, and often commutes along woodland rides. Maternity colonies are relatively large, with 20–200 individuals, of which up to a quarter may

WHERE TO LOOK/OBSERVATION TIPS

Confined to south-west England and Wales. A visit to a known roosting or emergence site with licensed bat workers offers the best chance of seeing this species.

See also: Horseshoe bats compared *p. 143*

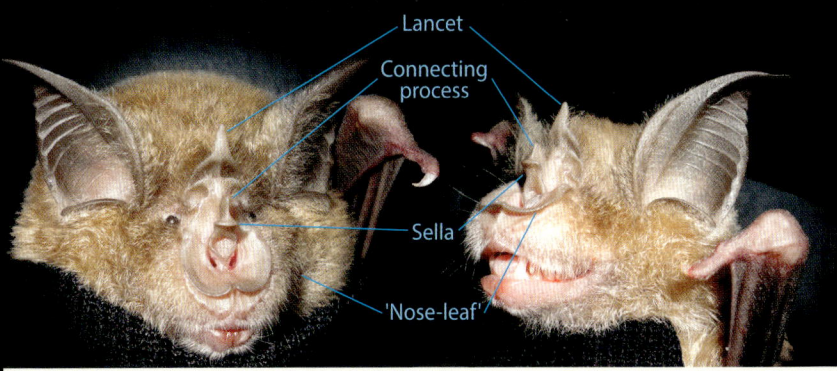

Lancet

Connecting process

Sella

'Nose-leaf'

Detail of Greater Horseshoe Bat head showing the structures unique to both horseshoe bat species.

Frequently forages low down, often close to vegetation or walls.

be males (unusual in bats), forming dense clusters. Sedentary; in summer they forage up to 2·5 km from the roost, and rarely travel more than 30 km between summer and winter roosts.

Breeding behaviour: Males form mating roosts in roofs of buildings or underground sites from late summer onwards. Visiting females may copulate with the same males year after year at these traditional sites. Some mating also occurs in spring. The single pup is born between the end of June and the end of July.

Population and status: British population estimated at fewer than 6,600 individuals, with the first record for Ireland, a single male, in 2013 (Co. Wexford). The recent trend shows slight increases year-on-year. Previously much more abundant; the overall decline in the last 100 years could have been up to 90%.

Characteristically wraps wings around body when at rest.

Horseshoe bats compared

Greater Horseshoe Bat

FOREARM >53 mm
WINGSPAN 35–40 cm

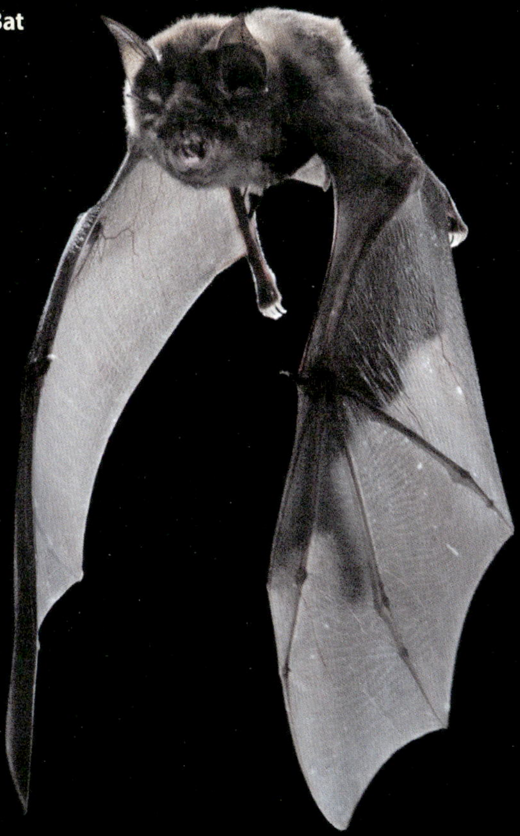

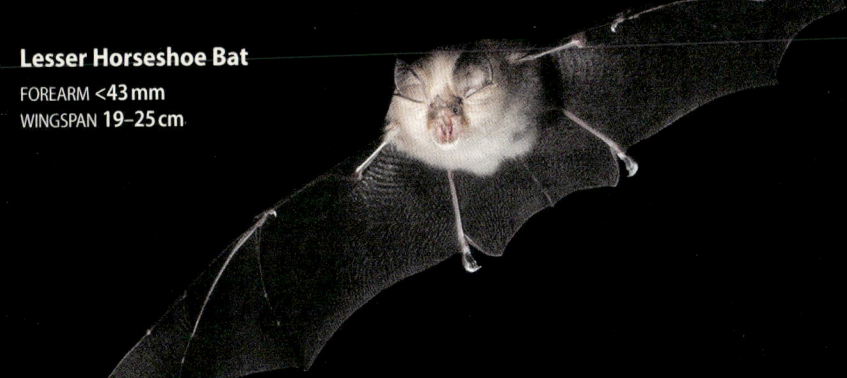

Lesser Horseshoe Bat

FOREARM <43 mm
WINGSPAN 19–25 cm

See also: Sounds *p. 136* | Droppings *p. 139*

Lesser Horseshoe Bat

Rhinolophus hipposideros

Little larger than a Common Pipistrelle (*page 154*), the delicate Lesser Horseshoe Bat has suffered drastic declines (by 90% or more) in many parts of northern Europe in recent times; in Britain, it disappeared from the north of England between the 1960s and 1990s. The declines have been attributed to toxic pesticides, especially in timber treatments. Now the issue has been recognised and is being addressed, it appears that a fragile recovery may be underway, but this bat is still rare.

LEGALLY PROTECTED	
Biodiversity List (En, Wa)	
Rare	
HB:	3·7–4·5 cm
WS:	19–25 cm
Wt:	4·0–9·5 g

Identification: A small bat with pale brown, soft, fluffy fur, rounded ears with a pointed tip and the unmistakable pink 'nose-leaf' typical of horseshoe bats. The young have greyish fur for their first year. Greater Horseshoe Bat (*page 140*) is very considerably larger (Lesser Horseshoe Bat is plum-sized; Greater Horseshoe Bat is pear-sized) and has paler brown wing membranes. When torpid at roost, both horseshoe bats wrap their wings around their body, and hang free.

Flight: Slow and fluttering with short glides, but very agile; faster wingbeats than Greater Horseshoe Bat. Often flies very low, gleaning prey from surfaces; sometimes circles closely around trees and shrubs to startle insects.

Sounds: A pulsed, high-pitched warbling at 112 kHz; pulses are shorter than for Greater Horseshoe Bat.

Signs: Summer roosts: Usually in the roofs of old buildings (*e.g.* churches), with direct, unhindered access into the roost. Generally hangs separately, not gathered into dense clusters. **Winter roosts:** Underground, in cellars, tunnels, caves, mines.

Habitat: Forages mainly inside deciduous woodland, but also uses tree and hedgerow lines. Does not normally fly over open ground.

Food: Small flies, such as craneflies, non-biting midges and mosquitoes, as well as beetles, spiders and moths. Catches food on the wing, by both aerial hawking and gleaning from vegetation *etc.*

Close-up of head showing 'nose-leaf'

Habits: Hibernates from November to March, but will feed during mild winter nights. Emerges 20–50 minutes after sunset, and typically hunts in woodland, especially in the shrub layer, up to 10 m above the ground. Nursery roosts contain 30–500 animals, including females, yearlings and often some males. Sedentary; wintering sites are usually within 5 km of maternity roosts.

Breeding behaviour: Mating takes place September–April, after a brief courtship chase. Maternity roosts form from April, each female giving birth to single pup between mid-June and mid-July. Young can fly at three weeks and are independent at six weeks.

Population and status: In Britain about 15,000 adults, increasing slightly; in Ireland approximately 12,500, stable or increasing.

WHERE TO LOOK/OBSERVATION TIPS

Easy to see at roost (*e.g.* when hibernating) or when emerging from a roost at dusk, but rarely seen at large, due to its scarcity and tendency to feed deep within woods.

See also: Horseshoe bats compared *p. 143*

Not much bigger than a pipistrelle, the Lesser Horseshoe Bat has slow, fluttering flight.

When roosting, hangs free on long legs, with wings wrapped around its body.

Serotine
Eptisecus serotinus

A large bat found mainly in southern England. The name derives from the Latin *serotinus*, meaning 'coming late', a reference to this species' relatively late evening emergence.

Northern Bat (*p. 186*)
Particoloured Bat (*p. 187*)
Big Brown Bat (*p. 194*)

LEGALLY PROTECTED (UK)

Uncommon

HB:	6·6–9·2 cm
WS:	32–38 cm
Wt:	15–39 g

Identification: Larger than most other regularly occurring British bats with a distinctive broad muzzle, relatively long, erect ears, and blackish face. The tail extends at least 2 mm beyond tail membrane. The fur is long and silky, mid- to dark brown (sometimes with golden-brown tips) on the upperparts and clearly paler, often yellowish-brown, on the underparts. Noctule (*page 148*) is larger and has short, rounded ears, while the structurally similar Common Pipistrelle (*page 154*) is much smaller, with less pointed ears and evenly coloured fur. The rare Northern Bat (*page 186*) is darker above and paler below.

Flight: Broader-winged and slower than Noctule, without the spectacular dives, and will forage closer to vegetation or the ground. It flies in broad circles over open areas or around trees, typically foraging 4–12 m above the ground.

Sounds: Loud 'smacks' with a distinctive irregular rhythm involving 3–5 calls per second, at a typical peak frequency of 26 kHz. Call character can overlap with Leisler's Bat (*page 152*).

Signs: Summer roosts: In older buildings, especially roof spaces. Often in houses with high gables. **Winter roosts:** In buildings; also occasionally caves.

Habitat: A broad range, including pastureland, woodland edge, parkland, hedgerows and suburban areas.

Food: Predominantly large beetles (especially dung beetles) and moths, and some flies and ichneumons, caught by aerial hawking. Prey is sometimes brushed from vegetation or captured on the ground.

Habits: Hibernates from November to the end of March/April. Emerges 20–25 minutes after sunset, much later than the Noctule, the other regularly occurring large British bat. Hunts in an area averaging 4·6 square kilometres, travelling between up to ten separate feeding sites along linear features (*e.g.* hedgerows). Maternity roosts form in May and comprise 10–60 adults, sometimes more. Most males roost singly in summer. Sedentary, with only small local movements.

Breeding behaviour: Mating takes place in September and October. The single pup is born in early July (although there are late births into August), flies at 4–5 weeks of age, and is usually weaned by 5–6 weeks.

Population and status: British population is about 15,000 (including 500 in Wales); appears to be stable and possibly increasing slightly. Absent from Ireland.

WHERE TO LOOK/OBSERVATION TIPS

Often forages, flying in wide circles, over parkland at rooftop height and also feeds around street lights: easy to confirm identification using a bat detector. Local bat groups monitor roosts and emergences.

Key morphological details

FOREARM: 48–58 mm. TRAGUS: longer than it is broad, one third the length of the ear and slightly curved forwards. TEETH: large. TAIL: has a slight post-calcarial lobe, and extends 2–6 mm beyond the tail membrane.

Has broad wings and a relatively slow, leisurely flight with short glides.

The contrastingly dark face and long, pointed ears and pointed tragus are distinctive, especially compared to other larger bats.

See also: Sounds *p. 136* | Droppings *p. 139*

Noctule
Nyctalus noctula

Several features make the Noctule easy to identify: it is the largest common British bat, which frequently flies at sunset or before, and typically hunts high over open areas, well above rooftop height.

LEGALLY PROTECTED (UK)
Biodiversity List (En, Wa, Sc)

Fairly common

HB:	6·0–8·9 cm
WS:	32–45 cm
Wt:	18–40 g

Identification: Gives the impression of large size, even without anything for comparison. Its short, smooth fur is a warm reddish-brown all over, and distinctly glossy, although duller after the moult in summer. The face is dog-like, with a dark muzzle, and it has widely spaced, rounded ears with a mushroom-shaped tragus. It also has a post-calcarial lobe. Leisler's Bat (*page 152*) is similar but smaller, and has distinctly two-toned fur on the upperparts: dark brown at the base with reddish-brown tips. The Serotine (*page 146*) is usually considerably darker and has much longer and more pointed ears and tragus. At dusk, could initially be mistaken for a flying bird but is easily distinguished by its wedge-shaped tail.

Flight: The long, narrow wings enable it to fly at 50 km/h, the impressive flight characterized by steep dives after large insects as the bat hunts over open areas. Typically flies at 10–50 m above the ground, but can go much higher.

Sounds: Loud, regular "*chip, chop*", the two calls with peak frequencies of 24 kHz and 19 kHz respectively, of 2–4 calls per second. Call character can overlap with Leisler's Bat. Some people (especially children) can hear the foraging or commuting calls without a bat detector; in addition, audible social calls can include a noisy chattering at roost.

Signs: Has a very noticeable musky, ammonia-like smell. Well-used roosts may have dark streak of urine and droppings below the exit hole. **Summer roosts:** Tree holes, especially those made by woodpeckers; also bat-boxes. **Winter roosts:** Tree holes and bat-boxes.

Habitat: A tree bat as far as roosting is concerned, but can be seen feeding over a range of habitats, including suburban areas, water bodies and woodland edges.

Food: A variety of small to large insects, including beetles, moths and flies, caught by aerial hawking, sometimes in daylight.

Habits: Hibernates from December to March. In summer, the earliest to emerge of all British bat species, often well before sunset, and feeds for about an hour before resting. Has up to six feeding bouts a night (more for lactating females), with a final burst of activity at dawn. Nursery roosts are of modest size, usually 20–60 females, separate from male summer roosts of up to 20 individuals. In warm conditions roosting animals are often noisy, and audible from a distance. Winter roosts hold both sexes. Noctules are migratory in mainland Europe, but there is no strong evidence that British animals move long distances.

WHERE TO LOOK/OBSERVATION TIPS

Choose a warm, still evening in summer almost anywhere within its range, and a clear view of the sky. The first bat to appear is likely to be a Noctule, flying high.

The largest common British bat, the Noctule has long, pointed wings and flies high and fast, with frequent swoops and steep dives after flying insects.

Mushroom-shaped tragus – a feature shared only with the smaller Leisler's Bat.

Breeding behaviour: Males establish mating roosts from early August, and sing to attract females, which also respond to the smell of secretions around the roost entrance. Roosts are defended against other males; the occupant acquires a harem of 4–5 females, mating taking place into the autumn. The female gives birth to a single pup in June or July, suckling it for about six weeks; she may leave it in a crèche while hunting. The young are independent at 6–7 weeks of age.

Population and status: British population estimated at 50,000, mostly in England and Wales. Although thought to be stable overall, it has declined in some areas, perhaps owing to intensive farming practices. Absent from Ireland.

Roosts in tree holes, particularly old woodpecker holes, and often emerges before sunset.

Noctule and Leisler's Bat compared

Noctule

HEAD–BODY LENGTH: **60–89 mm**
FOREARM: **47·3–58·9 mm**
WINGSPAN: 32–45 cm
FUR: short, smooth and distinctly glossy (duller after summer moult); warm reddish-brown all over
EAR: uniformly dark

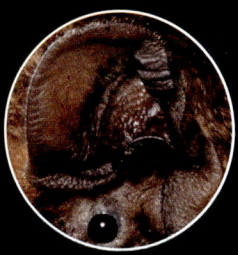

Leisler's Bat

HEAD–BODY LENGTH: **48–72 mm**
FOREARM: **38·0–47·1 mm**
WINGSPAN: 26–34 cm
FUR: longer than Noctule, especially thick around the neck; two-toned – dark brown at the base with reddish-brown tips; the fur on the underparts is usually paler than on the upperparts
EAR: base, and base of outer edge, usually pale

Leisler's Bat

Nyctalus leisleri

(Lesser Noctule)

Rare Britain;
common Ireland

HB:	4·8–7·2 cm
WS:	26–34 cm
Wt:	8–20 g

Closely related to the Noctule, Leisler's Bat similarly forages high in the air, flying fast and direct. Generally scarce, it is quite difficult to find in Britain, although relatively common in Ireland, where Noctule is absent.

Identification: A medium-sized bat with broad, rounded ears and a mushroom-shaped tragus, unlike those of other resident bats apart from the larger Noctule (*page 148*); it also has a post-calcarial lobe. The fur is bicoloured: the hairs on the upperparts being dark brown at the base with reddish-brown tips; on the underparts the fur is usually paler and has a yellowish wash. Compared with Noctule it also has longer fur, which is particularly thick around the neck and upper back, forming a mane, especially in males. The fur extends onto the wings and along the forearm, and the outer edge of the base of the ear is whitish – see *page 151* for comparison.

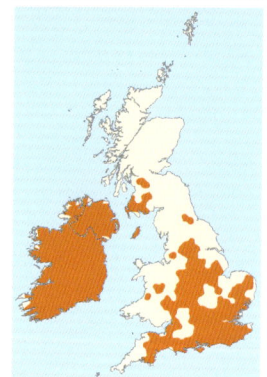

Flight: As with Noctule, flies high and direct on long, narrow wings, although its feeding swoops and dives are typically shallower.

Sounds: Similar to Noctule, but less regular: the "*chip*" and "*chop*" sounds are interwoven and not so strictly one after the other, given at a rate of 3–5 calls per second. The peak frequency of the "*chip*" call is around 27 kHz and the "*chop*" call is at 23 kHz (24 kHz and 19 kHz respectively in Noctule). Makes noisy social calls at roost which are often audible on warm days, as is the chirping territorial song of the male. Call character can overlap with Serotine and/or Noctule.

Signs: Summer roosts: Roof spaces of buildings; sometimes in tree holes and bat-boxes. **Winter roosts:** Bat-boxes and tree holes, often in Beech and oak trees. Also sometimes in buildings.

Habitat: Feeds mainly over pasture (especially with cattle), woodland edges, parkland and lakes/rivers. Does not usually feed over urban habitats, but will hawk around street lights.

Food: A range of small and medium-sized insects, including beetles, moths, flies (especially non-biting midges and Yellow Dung Fly) and caddisflies, caught by aerial hawking.

Habits: Hibernates from November to April. In summer it emerges early, 5–20 minutes after sunset. Nursery colonies are modest in size, just 20–50 individuals; summer roosts of males can reach ten or more. Changes roosts frequently in summer, even on a daily basis. Individuals travel up to 4 km to feed and do not hold territories. Migratory in mainland Europe; movements of up to 200 km have been recorded in Britain and Ireland.

Breeding behaviour: Mating takes place from the end of July to September or October. Each male defends a mating roost against other males, and can attract a harem of up to nine females. The territorial song of the male is audible to humans, and delivered from a perch or song-flight near the roost entrance. The single pup is born from mid-June and can fly after 30 days.

Population and status: Britain about 9,000; Ireland around 20,000.

WHERE TO LOOK/OBSERVATION TIPS

Visiting an emergence site or roost is the only realistic way of seeing this species in Britain or Ireland (where it is most numerous); usually seen in the early evening.

See also: Noctule and Leisler's Bat compared *p. 151*

Leisler's Bat is similar in shape to the Noctule in flight, with long, narrow wings.

Pipistrelles
3 regular British/Irish species, plus 2 vagrants

Small bats with widely spaced ears, a post-calcarial lobe, a rounded (but not mushroom-shaped) tragus, short forearm (<45 mm) and short tail extension beyond the tail membrane (<2 mm). The three regularly occurring species can be identified by their sounds (see *page 136*), but in the hand (which requires a licence) specific identification is based on certain morphological features: the extent and nature of the hairs on the tail membrane and tibia, details of the wing venation (though beware that this is not always foolproof and may be different on the two wings!), the actual and relative length of the fifth digit compared with the length of the forearm, and the colour of the penis. The two vagrant pipistrelles (*pages 188–189*) are identified on the basis of their dentition.

Common Pipistrelle
Pipistrellus pipistrellus

The most numerous and widespread bat in Britain and Ireland. A bat seen flying over a garden, or flitting through the car headlight beam, is most likely to be a Common Pipistrelle, although such views do not rule out the possibility of similar, less common species. Roosting mainly in buildings, it is a frequent occupant of bat-boxes, and mixes with other species when feeding over water or along woodland edges.

Identification: Britain's second smallest bat, only marginally larger than Soprano Pipistrelle (*page 158*). The fur is uniform brown (only slightly paler on the underparts) and the face and ears are contrastingly dark, often giving a 'bandit's mask' appearance. Soprano Pipistrelle, as well as having different echolocation calls, is slightly smaller, has a shorter muzzle, and pinker face and inside of the ear. Identification of Common Pipistrelle can be confirmed by the relatively sparsely haired tail membrane (the hairs extending only just beyond the body) and the wing venation (with a divided F 'cell' – see *opposite*). Additionally, the penis is grey-brown with a pale median stripe (see *page 157*), and the buccal glands are whitish. The smallest *Myotis* bats, such as the similarly-sized Whiskered Bat (*page 178*), have longer, more pointed ears with a pointed tragus, no post-calcarial lobe, and usually clearly paler fur on the underparts.

Flight: Fast-flapping and manoeuvrable, but typically makes sharper and less predictable twists and turns than many other species. Tends to fly at about 3–10 m above ground or water.

Sounds: When 'tuned-in' to peak frequency, a 'wet' smacking sound is heard at around 43–49 kHz, with a typical peak frequency at 45 kHz (55 kHz in Soprano Pipistrelle). In common with other bats, also makes social calls (*e.g.* a male defending territory, or individuals calling prior to departure from the roost), which may be heard as audible squeaks.

⟨◌⟩ Kuhl's Pipistrelle (*p. 188*)
Savi's Pipistrelle (*p. 189*)

LEGALLY PROTECTED
Biodiversity List (Wa, Sc)

Very common

HB:	3·5–4·5 cm
WS:	19–24 cm
Wt:	3–8 g

WHERE TO LOOK/OBSERVATION TIPS

Easy to see, except in the very north of Scotland (although beware confusion with Soprano Pipistrelle). Often seen by a lake or slow-flowing river at dusk, where it feeds with other species.

FOREARM: 29·2–33·5 mm. 5TH DIGIT: 37–41 mm. WING VENATION (from below): unbroken 'cell' between elbow and 5th digit (E), but 'cell' between centre of forearm and 5th digit (F) usually divided by a cross-vein (see *below*). BUCCAL GLANDS: whitish. PENIS: grey-brown with pale buff hairs and a distinctive pale median stripe. TAIL MEMBRANE: furred only close to the body.

DENTITION: as Soprano Pipistrelle; on upper jaw, I3 about half length of I2; P2 hidden behind canine; on lower jaw no gap between incisors (see *page 159*).

Often seen flying round houses and gardens

Pipistrelle wing venation
E 'cell' – the 'cell' between the elbow and the 5th digit
F 'cell' – the 'cell' between the middle of the forearm and the 5th digit

FOREARM

5TH DIGIT

ELBOW

unbroken + divided
E   F

'Bandit' mask; short ears with rounded tragus; no internarial ridge

Plagiopatagium cross-vein pattern unique to Common Pipistrelle and the rare Kuhl's Pipistrelle (*page 188*) in the region

Signs: Summer roosts: Usually in houses (including modern ones), often under roof tiles or felt, or in cavity walls. Also forms maternity roosts in tree holes and bat-boxes. **Winter roosts:** Roosts singly or in small groups in sites above ground, including crevices in buildings, tree holes and bat-boxes. Roosts are often marked by droppings adhering to walls, beams or windows, or on the ground below (see *page 138*). The pellets are black and cylindrical, 6–9 mm long.

Habitat: Occurs almost everywhere, although uncommon in city centres. It is the bat most likely to be seen in suburban gardens, but is also found in woodland, farmland and scrub, and over water.

Food: Small flying insects, particularly non-biting midges of the family Chironomidae, of which a single bat may consume 3,000 in a night. Obtains all its food by aerial hawking, catching prey with its tail membrane and feet.

Habits: Hibernates from November to April, but can sometimes be seen on a warm winter's night (temperature above 8°C) before returning to torpor. In summer, emerges about 20 minutes after sunset and feeds sporadically through the night. Feeding areas are usually close to the roost, on average just 1·5 km away, and individuals tend to patrol linear features on fixed flight paths. From May onwards, forms maternity roosts of 50–100 females, generally fewer individuals than roosts of Soprano Pipistrelle (*page 158*). Non-migratory; movements between summer and winter roosts are usually less than 20 km.

Breeding behaviour: As with all bats, mating takes place a long time before birth, between July and November. The sperm is stored inside the female's uterus until she ovulates in April. The single pup is born in mid-June in the maternity roost, becomes independent at four weeks old, and may breed the following summer. Females change their maternity roost sites frequently, sometimes to sites as far as 1·3 km away. Males begin attracting mates during the nursery period, when they establish roosts, often in bat-boxes, and perform song flights around them; if successful they can acquire up to ten females in a mating group.

Population and status: Abundant, British population estimated at more than 2·4 million; no estimates are available for Ireland.

Common Pipistrelle typically has a dark face ('bandit mask') and uniform fur.

A licence is required to handle bats	Nathusius Pipistrelle (*page 160*)	Common Pipistrelle (*page 154*)	Soprano Pipistrelle (*page 158*)
BODY FUR:	reddish brown (greyer in late summer) and shaggy on back, contrasting strongly with paler fur below	uniform brown above, only slightly paler below	pale reddish-brown above, slightly paler below
FACE/EARS:	pale to dark brown; internarial ridge absent	usually black, **darker and strongly contrasting** (**'bandit mask'**) **with the fur**; internarial ridge absent	pinker face and inside of ears – **never darker than fur**; internarial ridge usually present
TAIL FUR:	hair on **upper and lower surfaces dense and shaggy and to halfway down tail** and along tibia	hair on upper surface of tail membrane **extends just beyond body**, and on underside not along tibia	hair on upper surface of tail membrane dense but **extends no more than one third down tail**, and on underside not usually along tibia
WING: – see *p. 155* for definitions	E 'cell' **divided by a distinct bar** – see *p. 161*	E 'cell' unbroken – see *p. 155*	
		F 'cell' **usually divided by cross vein** – see *p. 155*	F 'cell' **undivided** – see *p. 159*
5TH DIGIT | FOREARM LENGTH (5TH DIGIT/ FOREARM RATIO):	41–48 mm* | 32·2–37·1 mm (>**1·25**) *usually >43 mm	37–41 mm | 29·2–33·5 mm (<1·25)	33–40 mm | 27·7–32·3 mm (<1·25)
PENIS:	ochre-brown with fringe of white hairs; never with pale median stripe 	grey-brown with pale buff hairs and pale median stripe	yellowish with yellowish hairs; never with pale median stripe 
DENTITION:	I3 usually projects above the short point of I2; P2 clearly visible beside canine – see *p. 161*	I3 about half length of I2; P2 hidden behind canine – see *p. 155* and *p. 159*	
BUCCAL GLANDS:	whitish		yellowish-brown to orange
CALL RANGE [PEAK]:	36–[**39**]–41 kHz	43–[**45**]–49 kHz	52–[**55**]–64 kHz

NATHUSIUS' PIPISTRELLE   COMMON PIPISTRELLE   SOPRANO PIPISTRELLE

See also: Sounds *p. 136* | Droppings *p. 139*

Soprano Pipistrelle

Pipistrellus pygmaeus

In the 1980s it became clear that the widespread pipistrelles of Britain and Ireland, known scientifically since the 1770s, were in fact two very similar species – Common Pipistrelle and Soprano Pipistrelle. Confirmed by genetic analysis, the two species differ most obviously in the pitch of their calls, but also in a variety of subtle morphological and ecological ways. A bat detector remains the simplest way to be certain as to which of the two species is present.

Kuhl's Pipistrelle (*p. 188*) Savi's Pipistrelle (*p. 189*)	
LEGALLY PROTECTED	
Biodiversity List (En, Wa, Sc, NI)	
Common	

HB:	3·5–4·5 cm
WS:	19–23 cm
Wt:	3–8 g

Identification: The smallest British bat, very similar to Common Pipistrelle (*page 154*) but with different echolocation calls. The facial skin tends to be paler than in Common Pipistrelle, and certainly not darker than the fur, which is pale reddish-brown, slightly paler below. Muzzle short, and there is usually a raised ridge between the nostrils (internarial ridge). In the hand, identification can be confirmed by the fairly hairy tail membrane (although the hairs extend no more than one-third of the length of the membrane and not usually along the tibia) and the wing venation (with both the E 'cell' and F 'cell' undivided – see *opposite*). The penis is uniformly yellowish (see *page 157*) and the buccal glands are yellowish-brown to orange. Soprano Pipistrelles are also smellier than Common Pipistrelles, the males especially giving off an obvious musky scent.

Flight: Rapid, agile and low (3–10 m), as Common Pipistrelle.

Sounds: Makes similar sounds to Common Pipistrelle, but the constant-frequency (CF) part of the call (the 'wet' sound) is typically at 51–60 kHz, with peak frequency at about 55 kHz.

Signs: As Common Pipistrelle, although roosts often have a musky odour.
Summer roosts: Maternity roosts in buildings (old and modern), *e.g.* behind tiles and in cavity walls; also tree holes and bat-boxes. **Winter roosts:** Crevices in trees and buildings; bat-boxes.

Habitat: Often associated with wetland habitats (more so than Common Pipistrelle), but by no means exclusive, and can be seen over gardens and woodland edges. Feeds less often over arable land and grassland than Common Pipistrelle.

Food: Small flies, including midges and mosquitoes. Believed to have a more specialised diet than Common Pipistrelle, taking more flies associated with aquatic habitats.

Habits: Hibernates September–April. In summer, emerges about 20 minutes after sunset, similar to Common Pipistrelle, but often commutes farther to feed (average 1·7 km). Maternity roosts are typically larger than those of Common Pipistrelle, from 20 to 800 individuals.

Breeding behaviour: Males occupy territories from the end of July, using social calls to attract up to 12 females; mating occurs mainly August–September. Females occupy maternity roosts from April and give birth to 1–2 pups from early June.

Population and status: British and Irish population estimated at 1·3 million, but should be treated as provisional given the problem of distinguishing this species from Common Pipistrelle.

WHERE TO LOOK/OBSERVATION TIPS

A widespread species, readily identified in summer using a bat detector. Look for it especially over lakes, rivers and small water bodies where it often forages under overhanging branches.

FOREARM: 27·7–32·3 mm. 5TH DIGIT: 33–40 mm. WING VENATION (seen from below): unbroken 'cells' between elbow and fifth digit (E), and between centre of forearm and fifth digit (F) (see *below*). NOSE: internarial ridge usually present. BUCCAL GLANDS: yellowish-brown to orange. PENIS: yellowish with yellowish hairs (see *page 157*). TAIL MEMBRANE: hairy to one-third its length.

DENTITION: as Common Pipistrelle; on upper jaw, I3 about half length of I2; P2 hidden behind canine; on lower jaw no gap between incisors (see *below*).

Soprano Pipistrelle is often associated with wetland areas and may commute farther from the roost to feed than Common Pipistrelle.

No gap between lower incisors

Face paler than Common Pipistrelle; never darker than fur; internarial ridge present

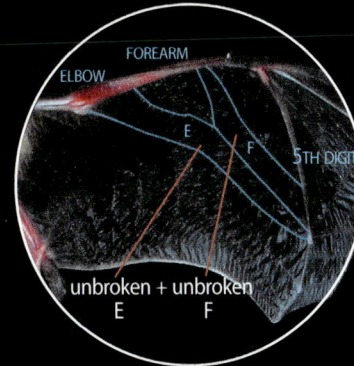

Plagiopatagium cross-vein pattern unique to Soprano Pipistrelle in the region

FOREARM
ELBOW
E   F
5TH DIGIT
unbroken + unbroken
E   F

Nathusius' Pipistrelle
Pipistrellus nathusii

First recorded in 1940, although subsequent records and the discovery of breeding roosts in Lincolnshire (1996), Northern Ireland (1997), and other locations since, suggests there is a resident population, augmented by autumn and winter visitors. In 2013, a British-ringed individual, recovered in the Netherlands, had travelled 596 km across the North Sea.

Kuhl's Pipistrelle (*p. 188*)
Savi's Pipistrelle (*p. 189*)

LEGALLY PROTECTED
Biodiversity List (Sc, NI)

Rare

HB:	4·6–5·5 cm
WS:	22–25 cm
Wt:	6–15 g

Identification: A very small bat, only slightly larger than Common (*page 154*) and Soprano (*page 158*) Pipistrelles but with noticeably longer and shaggier fur on the back. The fur on the upperparts is reddish brown (greyer in late summer) and contrasts markedly with the paler underparts. Unlike other pipistrelles, the tail membrane is shaggily and densely furred on both the upper and lower surfaces for half its length, and the hairiness also extends along the tibia. Identification can be confirmed by the wing venation (with both the E 'cell' and F 'cell' divided – see *opposite*). Also, the length of the fifth digit is usually >43 mm and the ratio of the length of the fifth digit to the length of the forearm is >1·25 (less in both Common and Soprano Pipistrelles). The penis is ochre-brown (see *page 157*) with a fringe of white hairs, and the buccal glands are whitish.

Flight: Slightly longer and broader wings than other pipistrelles with deeper wingbeats and a faster, more direct flight. Typically hunts 4–15 m above ground or water.

Sounds: As Common and Soprano Pipistrelles, but the constant-frequency (CF) part of the call (the 'wet' sound) is lower at 36–41 kHz, with peak frequency at about 39 kHz (similar to the rare Kuhl's Pipistrelle (*page 188*)). The pulses are longer and sound 'heavier', while the repetition rate is slower and more regular. Social calls are high-pitched but partially audible squeaks.

Signs: Summer roosts: Maternity roosts have been found in cavity walls and under slates of traditional brick buildings. **Winter roosts:** Cavities in brickwork, tree holes and rocks, but rare.

Habitat: Woodland, parkland and farmland with larger lakes and rivers, over which it feeds.

Food: Medium-sized aquatic insects such as midges and caddisflies, caught by aerial hawking.

Habits: Hibernates from November to about April. In summer, emerges around 20 minutes after sunset. Forms maternity colonies of between 20 and 200 from May onwards, sometimes with other pipistrelle species. Highly migratory: the northern and eastern European population moves south-west in the autumn when some come to Britain/Ireland, with regular reports from offshore oil-rigs and ships. Although some return in spring, there does appear to be a resident population.

Breeding behaviour: Unlike most bats, regularly produces two or even three young, in late May or June. The young can fly at four weeks and are able to feed themselves two weeks later. Males roost singly in summer and, from July to September, sing from territories close to maternity roosts. This attracts females which form a harem, within which mating takes place until November.

Population and status: The British population is estimated at 16,000 individuals; there are small numbers in Ireland.

WHERE TO LOOK/OBSERVATION TIPS
Using a bat detector by a large lake may prove successful.

Key morphological details

FOREARM: 32·2–37·1 mm. 5TH DIGIT: 41–48 mm (usually >43 mm). RATIO OF LENGTH OF FIFTH DIGIT TO FOREARM >1·25 (less in other pipistrelles). WING VENATION (seen from below): E 'cell' (between elbow and fifth digit) divided by a distinct bar (see *below*). BUCCAL GLANDS: whitish. PENIS: ochre-brown with fringe of white hairs (see *page 157*).TAIL MEMBRANE: densely and shaggily hairy for half its length on upper and underside.

DENTITION: on upper jaw, I3 usually projects above the short point of I2; P2 clearly visible beside canine; on lower jaw, obvious gap between incisors (see *below*).

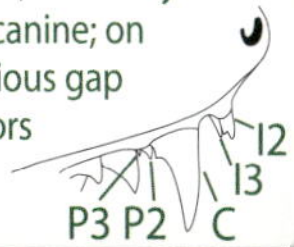

I2
I3
P3 P2
C

A wide-ranging, highly migratory bat.

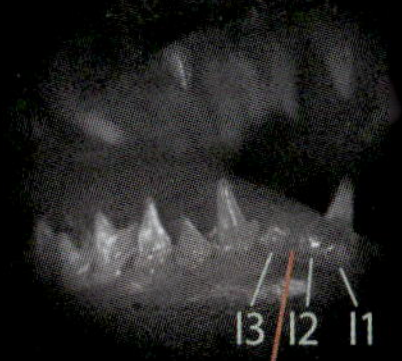

I3 I2 I1

obvious gap between lower incisors

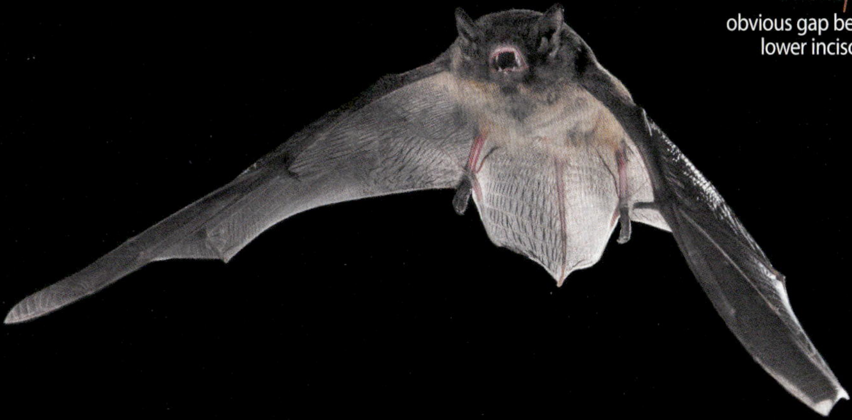

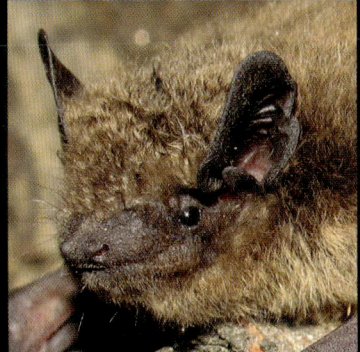

Face paler than Common Pipistrelle; dense, shaggy fur

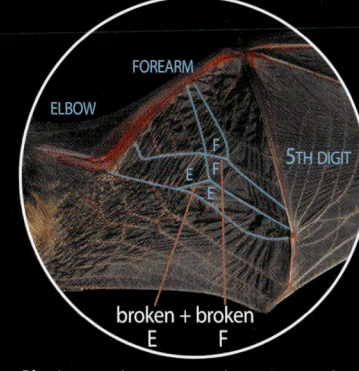

FOREARM
ELBOW
5TH DIGIT
F
F
E
E
broken + broken
E
F

Plagiopatagium cross-vein pattern unique to Nathusius' Pipistrelle in the region

NT # Barbastelle
Barbastella barbastellus

(Western Barbastelle)

A rare species of old woodland, the Barbastelle has both unusual roosting habits, behind peeling tree bark, and a unique combination of identification features, including black, silky fur and short, forward-facing ears.

LEGALLY PROTECTED (UK)
Biodiversity List (En, Wa)

Rare

HB:	4·0–5·2 cm
WS:	24–29 cm
Wt:	6–13 g

Identification: A distinctive medium-sized bat with a rather flattened face and broad, relatively short (<19 mm long), forward-facing ears that meet in the middle of the forehead. The ears often have a small flap of skin, generally referred to as a 'button' halfway along their outer edge. The bare parts, including the face and wing membranes, are very dark, and the fur is black, dense and long, sometimes with whitish tips on the back, giving a 'frosted' appearance.

Flight: Fast and agile when travelling, but slower, almost hovering, while foraging. Flies only after dark, typically up to 4 m above the crowns of trees.

Sounds: Sharp, clean "*tack*" calls of two types – louder calls with a peak frequency of about 32 kHz and quieter calls with a peak frequency of about 42 kHz, repeated at a rate of 8–9 calls per second; sounds like clacking castanets.

Signs: Summer roosts: Behind loose bark and in tree crevices; also in bat-boxes and sometimes in the roof spaces of buildings. **Winter roosts:** Sometimes behind bark, but also underground sites such as caves, mines and tunnels.

Habitat: Primarily a forest bat, largely dependent upon old and dead trees (especially oaks and Beech) and a structurally diverse understorey; a single colony may use up to 30 old trees with holes in the course of a single season. Often near rivers and ponds, especially when foraging, and will feed over water.

Food: Specializes on small moths, with some flies and beetles, mostly caught by aerial hawking.

Habits: Hibernates, but sometimes wakes up to forage during the winter months. In summer, typically emerges from about 30 minutes after sunset and for an hour or so thereafter. A wide-ranging bat, it often commutes some distance to feed at favoured foraging sites. Early evening feeding activity is below the canopy, moving higher as it gets darker. Maternity roosts are small, with fewer than 30 females, the location changing frequently, even on a daily basis. Males are solitary in the summer, as are both sexes in winter.

Breeding behaviour: Births occur from early July to August, each female rearing a single youngster. The young can fly at three weeks and hunt at six weeks of age. From late August, forms mating roosts comprising a single male with up to six females and their young.

Population and status: British population estimated at 5,000 individuals, found at low density. Absent from Ireland.

WHERE TO LOOK/OBSERVATION TIPS

Not likely to be encountered except on visits arranged especially to see this species, such as those run by local bat groups.

FOREARM: 36·5–43·5 mm. EAR: up to 18 mm long; broad, with 5–6 furrows and often a small 'button' halfway along outside edge. TRAGUS: broad-based, and strongly tapered towards the tip.

A medium-sized bat of old woodland, frequently feeding above the canopy.

Barbastelle has long, dark, silky fur, sometimes with 'frosty' tips to the hairs.

Long-eared bats
2 British/Irish species

The two long-eared bats can be readily distinguished by their extremely long ears, three-quarters the length of the body, that meet (or almost meet) on the forehead. However, specific identification can only be confirmed in the hand (which requires a licence) on the basis of certain morphological features: the length of the thumb and claws, the width of the tragus, the nature of the hairs on the feet and the shape of the penis.

Brown Long-eared Bat
Plecotus auritus

Widespread and relatively common: the third most abundant bat in Britain, after Common and Soprano Pipistrelles. With ears three-quarters the length of its body, it has an instantly recognisable profile both in flight and in the hand; at roost the ears are folded and tucked under the wings, leaving only the tragus exposed. Although it can echolocate, much of its food is detected audibly (*e.g.* the rustling of prey moving about on a leaf surface), or visually – its eyes are relatively large for a bat.

LEGALLY PROTECTED	
Biodiversity List (En, Wa, Sc, NI)	
Common	
HB:	4·5–6·0 cm
WS:	23–28 cm
Wt:	6–12 g

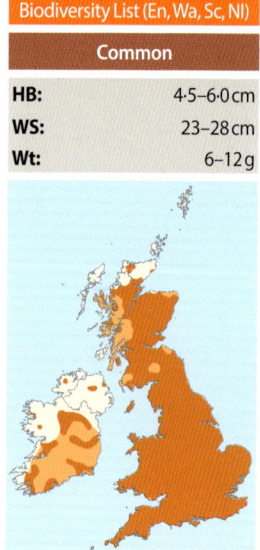

Identification: A small but distinctive bat, the very long ears distinguishing it from all but the very rare Grey Long-eared Bat (*page 168*); the rare Bechstein's Bat (*page 170*) has ears which are almost as long, but are well separated on the head, while those of long-eared bats are forward-facing and meet or almost meet across the forehead. Brown Long-eared Bat has long, brown to grey-brown fur on the upperparts, often tinged reddish, and slightly paler fur on the underparts. It is generally smaller than Grey Long-eared Bat and tends to have a shorter muzzle and paler face, but is only identifiable with certainty by its longer thumb and claws, narrower tragus, long hairs on the feet which curve away from the toes, and tapering, triangular-tipped penis (see *page 167*).

Flight: Slow and fluttering, usually very close to foliage, from ground level upwards. Broad wings give it agility, including the ability to hover close to leaves to pick prey off the surface.

Sounds: Often produces only quiet but rapid staccato clicks, which sound like a light purring, usually only audible within around 5 m. Echolocates through its mouth and nose. The call comprises two frequency-modulated sweeps, which together range from around 25–70 kHz, with a peak frequency at about 33 kHz (lower than typical frequency-modulated calls of *Myotis* bats).

Signs: Long-eared bats leave distinctive droppings and feeding signs, but these are probably not species-specific. **Droppings:** Medium to coarse-grained and 8–10 mm long; brown to black. **Feeding signs:** Moth wings and other debris often found below feeding perches, in outbuildings or porches. **Summer roosts:** Typically in buildings, including houses, churches and barns; less frequently in bat-boxes and tree holes. **Winter roosts:** Usually underground, in tunnels, cellars and mines, but also in tree holes and buildings.

WHERE TO LOOK/OBSERVATION TIPS

Many British woodlands support this species, and a summer night-walk with a strong torch could reveal one hunting in its distinctive way, identifiable by its unique silhouette. Often seen during checks of bat-boxes run by local bat groups (see *page 319*).

See also: Long-eared bats compared *p. 167*

Long-eared bats have a slow, fluttering flight on rather rounded wings.

The ears are three-quarters the length of the body.

Habitat: Principally found in deciduous woodland, but also in mixed woodland, villages and suburban areas.

Food: Takes a wide variety of invertebrates, particularly moths, large flies and beetles. Food is captured in two main ways: aerial hawking and foliage-gleaning. When gleaning, the bat often hovers briefly before snatching prey from a leaf surface. Larger prey is taken to a favoured perch, where it is eaten and the less palatable parts, such as wings, are dropped onto the ground below. Loses moisture easily from its ears and must drink regularly, taking gulps from the water's surface in flight.

Habits: Hibernates from November to the end of March/April, depending on the weather, but often wakes to feed when night-time temperature rises above 4°C. Emerges later in the evening than many other bats, from 40–60 minutes after sunset, sometimes earlier in the north. The first hunting session of the night lasts for about an hour. Rarely forages more than 1·5 km from roost; males typically travel farther than females. Highly sedentary, seldom moving far between summer and winter roosts; sometimes roosting in the same site throughout the year. Maternity roosts, occupied from May, comprise 10–50 closely related females and immatures that probably recognise one another by scent. Males occupy separate roosts until the young are weaned.

Breeding behaviour: Mating takes place in the autumn, and perhaps also in the spring. Swarming (see *page 37*) occurs from August to October at the entrances to caves some distance from the nursery roosts, thereby reducing the risk of inbreeding. There is also some swarming in spring. 1–2 pups are born in June; they can fly in six weeks and are weaned at 40–50 days.

Population and status: The British population is estimated at 245,000; thought to be stable at present, following a long-term decline brought about by changing farming practices, the conversion of barns to houses, and the use of pesticides in roofs where they roost. There are no estimates for Ireland, although it is widespread.

Long-eared bats partially roll up their ears when roosting.

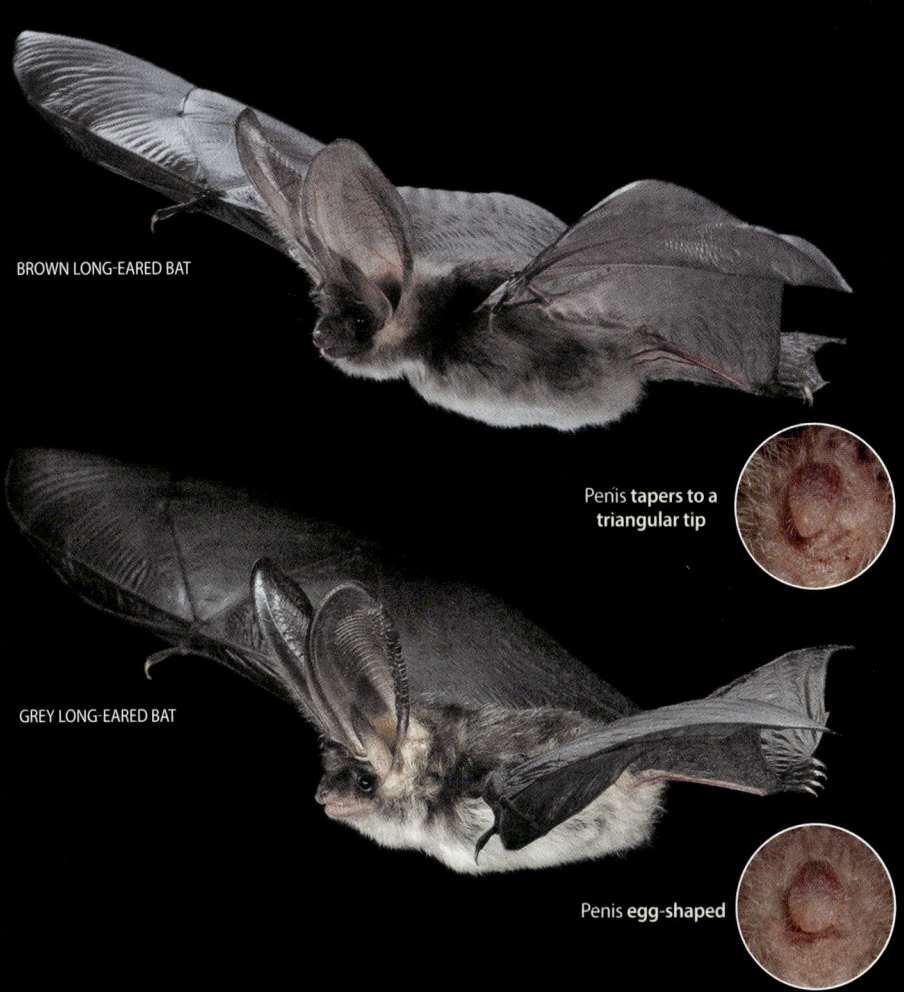

BROWN LONG-EARED BAT

Penis **tapers to a triangular tip**

GREY LONG-EARED BAT

Penis **egg-shaped**

Long-eared bats compared

	Brown Long-eared Bat *(page 164)*	Grey Long-eared Bat *(page 168)*
CALL RANGE [PEAK]:	25–[**33**]–70 kHz	17–[**33**]–56 kHz
HEAD–BODY LENGTH:	42–53 mm	41–58 mm
FOREARM:	35·5–42·8 mm	36·5–43·5 mm; ♂ usually >38 mm; ♀ usually >39 mm
THUMB:	**>6·5 mm long**	**<6·5 mm long**; rather short and thick
CLAWS:	**2·5–3·0 mm**	**<2·5 mm**
EAR:	>28 mm long	30–38 mm long
TRAGUS:	<5·5 mm wide; 14·0–16·7 mm long	>5·5 mm wide; 14·0–17·8 mm long
FOOT:	long hairs curving away from toes	toes with adpressed hairs

Grey Long-eared Bat   *Plecotus austriacus*

One of Britain's rarest breeding mammals, this bat is found primarily in the southern counties of England, from Devon and Somerset to West Sussex and the Isle of Wight. It has recently been discovered in Pembrokeshire, and has also been recorded in Norfolk and Leicestershire.

LEGALLY PROTECTED (UK)	
Very rare	
HB:	4·1–5·8 cm
WS:	25–30 cm
Wt:	7–14 g

Identification: The forward-facing ears of the Grey Long-eared Bat meet in the middle of the forehead and are remarkably long (three-quarters the length of the body), and distinguish this bat from all other British species except the Brown Long-eared Bat (*page 164*). Grey Long-eared Bat is, however, generally slightly larger and has darker, greyer fur on the upperparts, sometimes with a vague brownish tinge, and contrastingly paler fur on the underparts. It also has a longer muzzle and darker face. Identification needs to be confirmed by measuring the thumb and claw, both of which are shorter than on Brown Long-eared Bat, the width of the tragus (wider), and confirming the presence of adpressed hairs on the feet. The egg-shaped penis is also diagnostic (see *page 167*).

Flight: Similar to Brown Long-eared Bat: slow and fluttering, often low to the ground, close to and within vegetation; hovers frequently.

Sounds: As Brown Long-eared Bat, echolocation calls are usually very quiet and can be detected only within about 5 m. On a bat detector, rapid staccato clicks sound like very light purring across a broad range from 17–56 kHz, peaking at 33 kHz.

Signs: Indistinguishable from Brown Long-eared Bat. **Summer roosts:** In buildings, often in the attics of old houses and churches. **Winter roosts:** Underground, in cellars, tunnels and caves. Typically solitary or in single figures.

Habitat: Forages along woodland edges, in orchards and over meadows and pastures. Frequently in more open areas than Brown Long-eared Bat, and more strongly associated with human settlements, tending to avoid large woodlands.

Food: Moths (more than 70% of the prey), beetles, craneflies and other flying insects. Mostly caught on the wing, although will also glean from foliage.

Habits: Hibernates from October to March/April. Emerges late, between 40 and 60 minutes after sunset, and forages up to 5·5 km from the roost (farther than Brown Long-eared Bat). Maternity roosts are small, just 10–30 females, which may hang in groups or separately. Sedentary, moving only short distances between summer and winter roosts.

Breeding behaviour: Mating takes place in the autumn. Maternity roosts form in the late spring and the single pup is born in the second half of June. The young fly by August.

Population and status: The British population of this rare resident is estimated to be 1,000–1,500 individuals, although recent studies suggest a continuing decline. Absent from Ireland.

WHERE TO LOOK/OBSERVATION TIPS

Very difficult to see in Britain, the best chance being a visit to a roost with licensed bat workers.

See also: Long-eared bats compared *p. 167*

The muzzle is slightly longer than in Brown Long-eared Bat, giving a different profile.The fur on the upperparts is often greyer and darker than that of Brown Long-eared Bat, but this is not always a reliable identification feature.

NT Bechstein's Bat   *Myotis bechsteinii*

A rare and little-known British breeding species, Bechstein's Bat is found in just a few southern counties. Its ears are longer than all but the long-eared bats, to which it has a similar foraging style, using hearing as much as echolocation to find food. Roosts without curling up its ears, and shows swarming activity in the autumn.

LEGALLY PROTECTED (UK)	
Biodiversity List (En, Wa)	
Rare	
HB:	4·3–5·0 cm
WS:	25–30 cm
Wt:	7–13 g

Identification: A medium-sized bat with very long ears, a bright pink face, a pink tinge to the skin, and long fur which is mid-brown on the upperparts and whitish on the underparts. The very long ears can be almost as long as those of a long-eared bat (*pages 164–169*), but are widely spaced, not meeting in the middle of the forehead. When pushed forward the ears extend well beyond the tip of the nose, unlike in the closely related Natterer's Bat (*page 172*) which has shorter ears (extending only to the tip of the nose).

Flight: Slow and fluttering, although very agile, with short, broad wings, staying among trees.

Sounds: Similar to Whiskered Bat (*page 178*) and other *Myotis* species, but usually quieter, detectable from only a few metres away. Calls descend from around 116 kHz to 33 kHz in about 2·4 ms and have a peak frequency at 51 kHz, audible throughout as 'dry' clicks.

Signs: Summer roosts: In holes and fissures in old trees, or in bat-boxes. **Winter roosts:** Usually singly, in tree holes.

Habitat: A Beech forest specialist in mainland Europe, but in Britain usually found in oak woods with many veteran trees. Forages under dark, closed canopy, particularly in woodland with a well-developed understorey and close to water.

Food: Mainly moths, beetles, craneflies and lacewings caught by aerial hawking. Also gleans spiders and caterpillars from vegetation: these are detected by the sounds they make as they move around, rather than by echolocation.

Habits: Hibernates from October to April. A secretive bat, which in summer emerges from roost around 20 minutes after dusk, feeding close by and remaining under the canopy. Forms maternity colonies of 10–50 related adult females, although every few days these divide into groups in separate roosts with a changing membership. Males roost alone in summer. It is non-migratory, travelling only a few kilometres between the summer and winter roosts.

Breeding behaviour: Males and females meet at swarming sites (see *page 37*) in autumn, where it is assumed they mate. Maternity roosts form from early April; the single pups are born from late June to mid-July.

Population and status: The British population comprises about 1,500 animals; trend unknown. Absent from Ireland.

WHERE TO LOOK/OBSERVATION TIPS

The best chance of seeing this bat is during summer bat-box checks in known areas with licensed bat workers. Hard to see away from known roost sites as it forages high in the canopy.

Key morphological details

FOREARM: 39·0–47·1 mm. EAR: 21–26 mm long, projects >8 mm beyond the tip of nose; outside margin has 9–11 horizontal creases. TRAGUS: long and spear-like, extending halfway along the ear. CALCAR: straight, extending to one third of margin of tail membrane.

The rare Bechstein's Bat has the longest ears of any British or Irish bat, apart from the two long-eared bats.

The long ears that do not meet in the middle of the forehead distinguish Bechstein's Bat from both species of long-eared bat; the ears are longer than those of the other Myotis bats.

Natterer's Bat   *Myotis nattereri*

Generally uncommon, and only emerging when completely dark, this is a difficult species to see in Britain. With broad wings giving manoeuvrability, it hunts in confined spaces, from the canopy of trees down to just above the ground or water surface. It also gleans some prey from leaves and bark.

	Geoffroy's Bat *(p. 192)*
LEGALLY PROTECTED	
Biodiversity List (Sc)	

Locally common Britain; rare Ireland	
HB:	4·1–5·0 cm
WS:	24–30 cm
Wt:	7–12 g

Identification: A medium-sized bat with narrow, moderately long ears (extending to the tip of the nose when pushed forward), a unique 'S'-shaped calcar and a fringe of short, stiff hairs on the edge of the tail membrane. The fur on the underparts is white or buff, contrasting sharply with the brown fur on the upperparts, although juveniles are greyish-brown for their first year. The narrow muzzle and bare face, including around the eyes, are conspicuously pink, and the wing membrane also has a pinkish tinge. The widely separated ears and peaked muzzle give a dog-like impression to the face. Shares many features with Bechstein's Bat (*page 170*) but has shorter ears, while Daubenton's Bat (*page 174*) has darker fur on the underparts larger feet.

Flight: The wings are broad, and the flight relatively slow and agile, sometimes hovering. It has a tight turning circle when flying, and often flies among trees, foraging very close to the vegetation.

Sounds: A characteristically broad frequency range, from 107 kHz down to 23 kHz in about 4·7 ms, with peak frequency at 47 kHz. The pulses are very short, producing a rapid crackle on a bat detector. When hunting in tight spaces, the clicks change abruptly from slow to fast and back again.

Signs: Summer roosts: In natural tree holes, bat-boxes, cavities in buildings (especially large buildings with timber beams, such as churches and manor houses) and under bridges. **Winter roosts:** Usually underground, in caves and tunnels.

Habitat: Typically forages around woodland edges and water bodies such as rivers and lakes. Less frequently over arable farmland and pasture, but regularly visits cowsheds to hunt flies.

Food: A wide variety of invertebrates, with flies, caddisflies, bees, wasps and spiders being particularly favoured. Prey is caught in flight, gleaned from leaves and bark, or sometimes picked up from the ground after the bat has landed and crawled towards it.

Habits: Hibernates from November to April. In summer, emerges late in the evening, about one hour after sunset. Emergence is slow, just one at a time, never a cascade of bats as in, for example, the pipistrelles (*pages 154–161*). Maternity roosts comprise 25–200 adult females; in summer, males roost singly or in small groups. Largely sedentary, but may travel up to 120 km to winter roost.

Breeding behaviour: Mating is usually in October at traditional swarming sites (see *page 37*), such as cave entrances. Copulation may also occur in winter roosts. Maternity colonies form from mid-April; the single pup is born in June or early July, and can fly at about 20 days old.

Population and status: British population about 148,000; trend unknown. Rare in Ireland.

WHERE TO LOOK/OBSERVATION TIPS

Like Daubenton's Bat, frequently found over water, including small waterbodies with overhanging trees, but flies higher (over 1 m), rather than constantly skimming the surface.

Key morphological details

FOREARM: 34·4–44·0 mm. EAR: long; when folded forward projects just beyond the tip of nose; outer edge has 5–6 transverse folds. TRAGUS: longer than in similar species (over half the ear length) and more sharply pointed. CALCAR: uniquely 'S'-shaped, extending more than halfway along margin of tail membrane. TAIL MEMBRANE: stiff (1 mm long) bristles along edge.

The long, **'S'-shaped calcar** is unique to Natterer's Bat

Has a noticeably pink face and white fur on the underparts.

See also: Sounds *p. 137* | Droppings *p. 139*

Daubenton's Bat
Myotis daubentonii

Daubenton's Bat is one of the easiest bats in Britain and Ireland to observe and identify. It has a distinctive way of feeding – flying very low over the surface of open, still water to snatch insects caught on the surface, or which are flying just above it. This is one of the species that shows swarming behaviour in the autumn.

◐ Pond Bat (*p. 193*) Little Brown Myotis (*p. 195*)	
LEGALLY PROTECTED	
Biodiversity List (Sc)	
Common	

HB:	4·3–5·5 cm
WS:	24–27 cm
Wt:	6–12 g

Identification: A small bat, barely larger than the pipistrelles (*pages 154–161*) with which it is often seen hunting over water. In the hand, it is distinguishable from pipistrelles by its bare, pink face, relatively large feet, and a longer, blunt-tipped tragus that curves forwards slightly and extends half the length of the ear. The fur is relatively short, dense and shiny, the belly being off-white and contrasting with a dark brown back. Each hair is uniformly coloured from base to tip. Juveniles have darker faces than adults, with a well-defined black-blue mark on the lower lip. Daubenton's Bat has shorter and blunter ears than Natterer's (*page 172*), Whiskered (*page 178*), Alcathoe (*page 180*) and Brandt's (*page 182*) Bats, and a straight calcar that extends for about three-quarters of the length of the tail membrane. The hairs along the edge of the tail membrane are soft.

Flight: Typically flies very low, 5–25 cm above a still water surface, with fast, whirring wingbeats. Steadier flight than Natterer's Bat or the pipistrelles, often making long sweeping turns while other bats make sharper changes of direction.

Sounds: Calls fall from 81 kHz to 29 kHz in about 3·2 ms, with a peak frequency at 47 kHz. The clicks are staccato and fast, sounding like a distant machine gun.

Signs: Summer roosts: In holes in trees (usually long, vertical crevices) and, especially with males, bridges; only infrequently found in bat-boxes and buildings. Usually near water. **Winter roosts:** Underground sites such as caves, tunnels and cellars. Often singly in cracks, and hard to locate.

Habitat: Typically associated with still or slow-flowing freshwater for foraging, but will hunt along woodland edges and over meadows.

Food: Mostly small insects: non-biting midges (Chironomidae), mosquitoes, aphids, mayflies, craneflies, caddisflies and lacewings. Prey is initially located by echolocation and usually caught by being scooped from the water's surface using the tail membrane and feet. The bat then rolls forward to take the prey. A visible ripple is often left where a bat has caught prey.

Habits: Hibernates from at least December to April, but sometimes emerges to feed in mild weather. In summer, emerges late, often 45 minutes or more after sunset, and uses linear features such as hedgerows on the flight to a foraging site. Individuals spread out over the water and may

WHERE TO LOOK/OBSERVATION TIPS

One of the easiest bats to see and identify. Simply pick a mild summer's evening, find the nearest pond, lake or canal and look out for bats skimming very low over the water. On a heterodyne detector, the rhythm is like that of a purring cat.

FOREARM: 33·1–42·0 mm. EAR: relatively short; just reaching tip of nose. TRAGUS: long and straight or slightly curved, extending halfway up the ear. CALCAR: **straight, extending more than halfway (usually three-quarters) along margin of tail membrane**. TAIL MEMBRANE: soft hairs along edge. FEET: **large, longer than half the length of the tibia**, with bristly hairs on the toes.

Catches much of its food from at or near the water surface, using its large feet to grab prey and the tail membrane as a scoop.

Has a distinctive patch of bare pink skin around the eyes and short, rounded ears with a long, usually straight tragus; the short, shiny fur is clearly paler on the underparts.

briefly hold territories. Maternity roosts comprise 20–50 females; unusually, similar-sized roosts of males can also be found in the summer. In August and September, both sexes (but higher numbers of males) swarm (see *page 37*) at the entrances of caves or mines.

Breeding behaviour: Mating takes place from October (at swarming sites) right through into the winter, when the bats are hibernating. The single pup is born the following summer, from mid-June onwards; able to fly at three weeks, young can hunt for themselves when four weeks old.

Population and status: The British population is estimated at 150,000; the species is also widespread and numerous in Ireland although no population figures are available.

Daubenton's Bat has large feet and a long, straight calcar that extends for about three-quarters of the length of the tail membrane.

Whiskered, Alcathoe and Brandt's Bats (the 'WAB' group) compared

Three of the regularly occurring small *Myotis* species, the so-called 'WAB' group of Whiskered, Alcathoe and Brandt's Bats, provide a particularly difficult identification challenge. As a group, they can be told by their lack of a post-calcarial lobe (confirm this by checking for the absence of a 'T'-bar), relatively short ears (<18mm) which when pushed forward fall short of the tip of the nose, and straight calcar that extends for less than half the length of the edge of the tail membrane. The calls of these species are difficult to detect and are very similar, and consequently of little value for specific identification. Examination in the hand is, therefore, the only way of identifying these species, although in some cases (particularly with immature individuals) DNA analysis is required for confirmation.

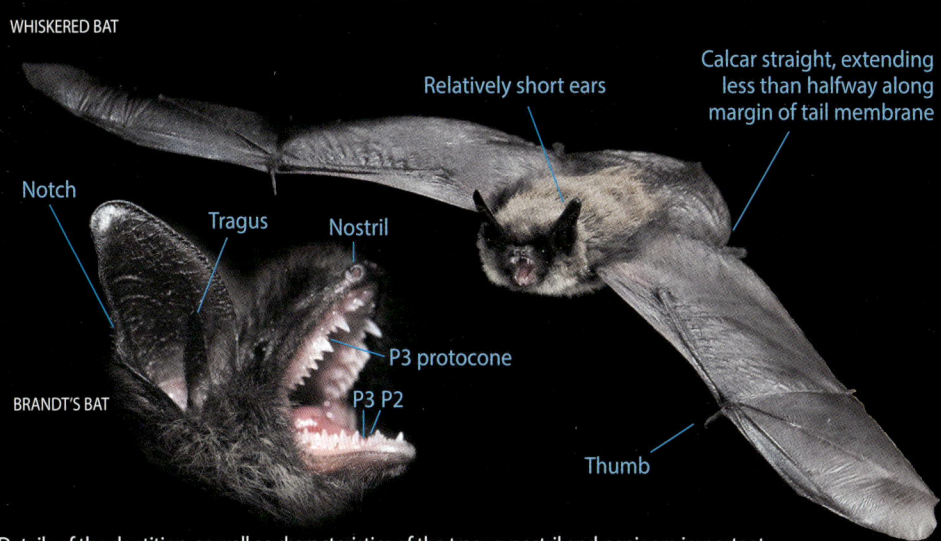

WHISKERED BAT

Relatively short ears

Calcar straight, extending less than halfway along margin of tail membrane

Notch

Tragus

Nostril

P3 protocone

BRANDT'S BAT

P3 P2

Thumb

Details of the dentition, as well as characteristics of the tragus, nostril and penis are important for identification – see *page 132* for comparative illustrations of these features.

	Whiskered Bat (*page 178*)	Alcathoe Bat (*page 180*)	Brandt's Bat (*page 182*)
FOREARM:	32·0–36·5 mm	29·7–34·8mm (usually <32·8mm)	33·0–38·2 mm
THUMB:	4·3–5·9 mm	3·8–5·0 mm	approx. 4·1–5·2 mm
NOSTRILS:	**round/comma-shaped**	heart-shaped	usually heart-shaped
TRAGUS: (see accounts and *page 132*)	dark; long, pointed, **slightly concave and curving forward**; extends at least as far as notch on outer edge of ear	**pale, particularly towards base; relatively short**; does not extend beyond notch on outer edge of ear	dark; long, pointed, **straight**; extends at least as far as notch on outer edge of ear
DENTITION: LOWER JAW 3RD PREMOLAR (P3):	third premolar (P3) less than half the size of second premolar (P2)		**third premolar (P3) almost as large as second premolar (P2)**
UPPER JAW P3 PROTOCONE:	**very small or absent**	pronounced	
PENIS: (see accounts and *page 132*)	parallel-sided for whole length		**distinctly swollen-tipped** (NB may be parallel-sided in younger animals)
	rather round-tipped	rather square-tipped	
FOOT LENGTH:	5·8–7·4 mm	4·2–5·8 mm (usually <5·6mm)	approx. 5·5–7·8 mm

Whiskered Bat

Myotis mystacinus

Whiskered Bat is so similar to Brandt's and Alcathoe Bats that much of what is known about this group of species is clouded by past identification issues. However, although uncommon, Whiskered Bat is widespread across England, Wales and Ireland, with a few records from southern Scotland.

Alcathoe Bat (*p. 180*)

LEGALLY PROTECTED
Biodiversity List (Sc)

Uncommon

HB:	3·5–4·8 cm
WS:	19·0–22·5 cm
Wt:	4·0–8·0 g

Identification: Very small, barely larger than a pipistrelle (*pages 154–161*), but with longer, more pointed ears, a black, bristly-looking face, and pale underparts which contrast with the dark brown upperparts. The fur on the upperparts is shaggy which, together with this bat's much smaller feet, helps distinguish it from the otherwise similar Daubenton's Bat (*page 174*). Whiskered, and the extremely similar Alcathoe (*page 180*) and Brandt's (*page 182*) Bats – the so-called 'WAB' group (see *page 177*) – can be separated from other small *Myotis* species by the relatively short calcar, which does not extend more than halfway along the edge of the tail membrane. However, specific identification within the 'WAB' group relies on a combination of characters, but even so DNA analysis may be required for confirmation.

Flight: Fast and fluttering with brief glides, like a Peacock butterfly in flight. Often patrols regularly over a linear feature such as a forest ride or hedgerow, in largely level flight with an occasional dive.

Sounds: Indistinguishable from Brandt's Bat: rapid 'dry' clicks from 88 kHz to 32 kHz in about 4·2 ms; peak frequency at about 48 kHz.

For Britain, map shows combined distribution of Whiskered and Brandt's Bats

Habitat: A bat of habitat margins and mosaics, including woodland edges and rides, hedgerows, agricultural areas, rivers, parks and gardens.

Signs: Summer roosts: In buildings, especially under cladding and tiles; rarely in trees and bat-boxes. **Winter roosts:** Underground, in caves and mineshafts, usually close to the entrance; occasionally cellars. In more humid locations than those used by Brandt's Bats.

Food: A wide range of small flying insects, including flies and moths, taken by aerial hawking.

Habits: Hibernates from December to at least March, but sometimes active in winter, even during the day. During summer it emerges from roost around 30 minutes after dusk and feeds throughout the night. The summer (maternity) roosts comprise 20–60 or more females, sometimes sharing with pipistrelles or other species. Males roost alone in summer. Gathers outside caves and other underground sites to swarm in August–October. In contrast to other swarming species, in which males outnumber females, the sex ratio is similar. Largely sedentary, rarely moving over 50 km.

Breeding behaviour: Mating occurs from August into winter. Births occur from mid-June in favourable weather, with a single pup or very occasionally twins. Young are weaned after a month, and although they can breed the following year usually first do so when more than a year old.

Population and status: British population is estimated at 40,000. There are 24,000 in Northern Ireland but no estimates for the Republic of Ireland. The trend is unknown.

WHERE TO LOOK/OBSERVATION TIPS

Look out for a small bat seemingly plying a regular beat back and forth along a hedgerow. Otherwise, join bat-box or roost emergence monitoring with a local bat group.

FOREARM: 32·0–36·5 mm. THUMB: 4·3–5·9 mm. DENTITION: third premolar on lower jaw less than half the size of the second, and protocone (cusp) on anterior angle of third premolar on upper jaw very small or absent. TRAGUS: long, pointed; extends above notch on outer edge of ear. NOSTRILS: round/comma-shaped. PENIS: parallel-sided for whole length. CALCAR: straight, extending less than halfway along margin of tail membrane. FOOT: 5·8–7·4 mm long.

The fluttering flight, level with brief glides, has been likened to that of a Peacock butterfly.

nostril round/comma-shaped

slightly concave

P3 less than half size of P2 on lower jaw (as Alcathoe Bat, *p. 181*); **protocone (Pr) small or absent**

tragus **dark, long, slightly concave**; extends at least to notch

penis parallel-sided + round-tipped

Whiskered Bat is small with a very dark face and wing membranes, and the fur is shaggier than that of the similar Daubenton's Bat. The ears typically look narrow and pointed and the nostrils are round or comma-shaped.

DD **Alcathoe Bat**   *Myotis alcathoe*

The rare Alcathoe Bat was identified as occurring in Britain in 2009, from records in Yorkshire and Sussex; since then they have been found in these areas fairly regularly. However, as it was confirmed as a distinct species only in 2001 (from Greece and Hungary), it may have previously been overlooked in Britain; its status is therefore uncertain. The Sussex records are from swarming sites. Its name comes from a princess of Greek mythology who lived near gorges and streams, as the bat does today in parts of its range.

LEGALLY PROTECTED (UK)	
Rare	
HB:	3·9–4·4 cm
WS:	20 cm
Wt:	3·5–5·5 g

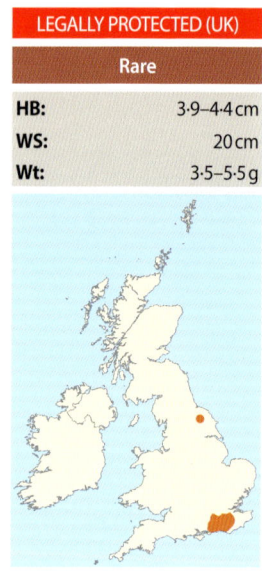

Identification: A very small bat, the smallest of its genus, easily distinguished from similar-sized pipistrelles (*pages 154–161*) by its longer, pointed ears, pointed tragus, blunt nose and paler fur on the underparts. The overall colour is similar to the much larger Daubenton's Bat (*page 174*). This is one of the so-called 'WAB' group (comprising Whiskered (*page 178*), Alcathoe and Brandt's (*page 182*) Bats) that is notoriously difficult to identify, but which can be separated from other small *Myotis* species by the relatively short calcar, which does not extend more than halfway along the edge of the tail membrane. Compared to Whiskered Bat it tends to have a reddish-brown tinge to the fur (adults only), paler ears and tragus, a shorter muzzle and paler pink face; Brandt's Bat is generally larger and has a darker face and golden-brown tips to the fur on the upperparts. However, specific identification can only be confirmed based on a combination of characters (see *page 177*), and even then DNA analysis may still be required to be certain.

Flight: Fluttering and highly agile; no known distinctions from Whiskered or Brandt's Bats.

Sounds: Generally indistinguishable from other *Myotis* bats with a heterodyne detector: frequency-modulated call drops from 120 kHz down to 43 kHz in <4 ms, with a peak frequency at 53 kHz. The high end frequency can be distinctive.

Signs: Summer roosts: In cracks in trees and behind peeling bark, often high in the canopy. **Winter roosts:** Recorded in caves; may also use trees and possibly Badger setts.

Habitat: Probably has specific requirements: in mainland Europe usually found in moist, mature deciduous forest near rivers or streams. It is likely to be dependent upon old and decaying trees for its roosts. Often forages at the edge of water.

Food: Mainly small flies, such as mosquitoes and gnats, caught by aerial hawking.

Habits: Still poorly understood. Hibernates from October to April. The few maternity roosts known in Europe are of modest size (maximum 80, usually fewer). Swarms in the autumn (from August), when bats congregate at cave entrances and fly around, probably as a prelude to mating.

Breeding behaviour: Little-known, but young have been found from the second half of June.

Population and status: Unknown, but probably rare.

WHERE TO LOOK/OBSERVATION TIPS

Few people have had the privilege of seeing an Alcathoe Bat in Britain; swarming sites in the south of England represent the best, albeit slim, chance for the would-be observer.

See also: Whiskered, Alcathoe and Brandt's Bats compared *p. 177* and *p.132*

Key morphological details

FOREARM: 29·7–34·8 mm (usually <32·8 mm). THUMB: 3·8–5·0 mm. DENTITION: third premolar on lower jaw less than half the size of the second, and protocone (cusp) on anterior angle of third premolar on upper jaw pronounced. TRAGUS: relatively short; does not extend beyond notch on outer edge of ear. NOSTRILS: heart-shaped. PENIS: parallel-sided for whole length. CALCAR: straight, extending less than halfway along margin of tail membrane. FOOT: 4·2–5·8 mm long (usually <5·6 mm).

A golden-brown tinge to the fur and pinkish face are indicative of Alcathoe Bat. Both the ears and tragus are relatively short.

nostril heart-shaped

Pr

P2 P3

P3 less than half size of P2; protocone (Pr) pronounced

tragus **pale, short, straight**; does not extend above notch

penis parallel-sided + square-tipped

Alcathoe Bat has a relatively short muzzle compared to Whiskered and Brandt's Bats.

Brandt's Bat

Myotis brandtii

Differentiated from Whiskered Bat as recently as 1970, Brandt's Bat is difficult to see and challenging to identify, even in the hand. It is a small bat with pointed ears, found in woods and forests, typically close to water.

Alcathoe Bat (*p. 180*)	
LEGALLY PROTECTED	
Biodiversity List (Sc)	
Uncommon	
HB:	3·9–5·1 cm
WS:	21–24 cm
Wt:	4·5–9·5 g

Identification: A small, dark bat with pointed ears and shaggy fur on the upperparts. Being one of the so-called 'WAB' group (comprising Whiskered (*page 178*), Alcathoe (*page 180*) and Brandt's Bats), identification is far from straightforward, but the relatively short calcar, which extends only up to halfway along the edge of the tail membrane rules out other small *Myotis* species. Brandt's Bat is similar to Daubenton's Bat (*page 174*) but has longer ears, a longer, more pointed tragus, a darker face, less smooth fur and relatively smaller feet. It is marginally larger than the other species in the 'WAB' group and the fur on the upperparts is sometimes golden-tipped in older animals (aged 7 years or more); the underparts are light grey with a yellowish tinge. However, specific identification requires a combination of characters (see *page 177*), and even then DNA analysis may be required for confirmation.

For Britain, map shows combined distribution of Brandt's and Whiskered Bats

Flight: Indistinguishable from Whiskered Bat: fast, fluttering and agile, with quick turns.

Sounds: Indistinguishable from Whiskered Bat: rapid 'dry' clicks down a wide range of frequencies from around 92 kHz to 34 kHz in about 3·5 ms, with peak frequency at about 47 kHz.

Signs: Summer roosts: Buildings, especially older ones with stone walls and slate roofs. Occasionally found in tree holes or bat-boxes. **Winter roosts:** Underground, in caves and mine-shafts; sometimes in cellars.

Habitat: Deciduous, mixed and coniferous woodlands with water nearby. Sometimes feeds along hedgerows, but more of a forest dweller than Whiskered Bat.

Food: Small flying insects, including midges and moths, taken by aerial hawking. Also some non-flying prey such as spiders, probably gleaned from foliage.

Habits: Hibernates from November to March; males may continue until May. In summer, emerges about 30 minutes after sunset and feeds intermittently all night. The maternity roosts comprise 20–60 females, sometimes sharing sites with other species, including Whiskered Bat. Males roost alone, and both sexes hibernate singly. Swarms in the autumn (see *page 37*). Non-migratory, moving only short distances between summer and winter haunts.

Breeding behaviour: Mates in autumn/winter, perhaps at swarming caves in August–October or at a winter roost. Gives birth in June (later in adverse weather); the single pup flies at three weeks, and is able to forage a week later.

Population and status: British population estimated at 30,000; first recorded in Ireland in 2003.

WHERE TO LOOK/OBSERVATION TIPS

The only realistic way to see this species in Britain is to contact a local bat group and join trips to monitor roosts or swarming sites.

The shaggy fur is light grey on the underparts.

nostril round/comma-shaped

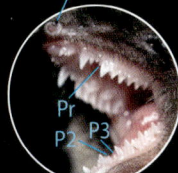

P3 almost same size as P2
protocone (Pr) pronounced

straight

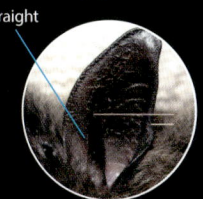

tragus **dark, long, straight**;
extends at least to notch

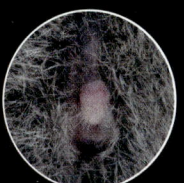

penis parallel-sided
+ swollen-tipped

Has long ears and long, pointed tragus, and usually heart-shaped nostrils.

Greater Mouse-eared Bat   *Myotis myotis*

For much of its recent history in Britain, this species has been represented by a single male hibernating at a roost in Sussex (2002–present). However, it appeared in other winter roosts regularly between 1958 and 1988, and potentially could do so again. A fairly common and widespread species in mainland Europe, it has the unusual habit of gleaning most of its food from the ground.

LEGALLY PROTECTED (UK)	
Very rare	
HB:	6·7–8·4 cm
WS:	36–45 cm
Wt:	28–40 g

Identification: The combination of large size – as big as a Noctule (*page 148*) – pale body (light brown above, whitish below) and long, broad ears with transverse folds and a pointed tragus makes the Greater Mouse-eared Bat unmistakable. Its face is pink, the fur is short and dense, and the nose is short and broad, while the wings are noticeably leathery.

Flight: Relatively slow and direct on long, broad wings, with a distinctive 'rowing' action. When foraging, it usually flies only 1–2 m above the ground, with head and ears facing downwards, hovering before a strike, in a manner similar to that of an owl.

Sounds: Frequency-modulated call falls from about 80 kHz down to 28 kHz in about 4·6 ms, with a peak frequency at 36 kHz. Loud and relatively slow.

Signs: Summer roosts: Usually in the roof spaces of old buildings. **Winter roosts:** Caves, tunnels and other underground sites, preferring relatively warm locations with a temperature of 6–12°C.

Habitat: In mainland Europe, essentially a forest bat, found in stands of both deciduous and coniferous trees, and mixtures of the two. Wherever it occurs it has a preference for forests with little or no ground cover and therefore easy access to ground-dwelling invertebrates. It also hunts over open areas, including pasture and meadows, and around rural settlements.

Food: Large insects, such as terrestrial beetles, grasshoppers and crickets, and caterpillars. It catches a large proportion of its prey on the ground, detecting it by smell as well as by hearing. Since echolocation does not work with food on the ground, this bat is thought to specialize on invertebrates that make loud rustling sounds when moving about in the leaf-litter.

Habits: Hibernates from November to March. When active, individuals emerge well after dark, and may travel up to 26 km to feed. It is one of the more migratory bat species, frequently moving 50–100 km between summer and winter roosts in mainland Europe.

Population and status: Declared extinct in Britain in 1990, after the last animal from the only known roost died. However, in recent years occasional individuals have been discovered elsewhere, perhaps suggesting that a colony survives, or that further vagrancy or colonization from elsewhere in Europe has occurred.

WHERE TO LOOK/OBSERVATION TIPS

Might yet be detected in forests in southern England, but for now only one individual is known.

FOREARM: 55·0–66·9 mm. EAR: >24·5 mm long and >16 mm wide, with 7–8 transverse folds.

Forages very low, just above the ground.

A pale bat, light brown above and clean white below

Northern Bat

Eptisecus nilssoni

LEGALLY PROTECTED (UK)	
Vagrant	
HB:	5·4–6·4 cm
WS:	24–28 cm
Wt:	17–34 g

A mainland European species and the only bat that occurs regularly inside the Arctic Circle. There are two records of presumed natural origin in Britain: one hibernating in Surrey (January 1987) and one on a North Sea oil-rig (August 1996).

Identification: A medium-sized bat, similar to Serotine (*page 146*) with black ears and face, but slightly smaller and has rounder ears. The fur on the upperparts is dark with golden tips, whereas on the underparts it is contrastingly yellowish-brown (the contrast between upper and underside is less obvious in Serotine). Particoloured Bat is superficially similar, but has silvery tips to the fur and is whiter on the underparts, as is the much smaller Savi's Pipistrelle (*page 189*).

Flight: A fast, agile flier at varying heights above ground.

Sounds: Similar to Serotine but with a strong constant-frequency 'tail' at 26–29 kHz, and peak frequency at 27 kHz.

Signs: Summer roosts: Buildings (*e.g.* under slate roofs and behind cladding). **Winter roosts:** Underground in mines and caves.

Habitat: Principally boreal and mountain forest, both coniferous and deciduous. Often feeds near lakes and streams, and flies around street lights.

Habits: Emerges early, sometimes by day; highly territorial when hunting. In mainland Europe tends to move south-west during the autumn.

Population and status: A rare vagrant: two wild records and two that were accidental imports on ships.

Key morphological details

FOREARM: 37–44 mm (Serotine: 48–58 mm). EAR: 5 transverse folds. TRAGUS: short and broad. PENIS: broadly oval (Particoloured Bat: long and thin). TAIL: last vertebra extends ≤4 mm beyond tail membrane.

Particicoloured Bat   *Vespertilio murinus*

LEGALLY PROTECTED (UK)	
Rare migrant	
HB:	5·4–6·6 cm
WS:	27–31 cm
Wt:	12–20 g

Although rare, this is one of the most frequent non-breeding migrant bats to reach Britain, and recently has become an almost annual visitor.

Identification: A medium-sized bat distinguished by the 'frosty' tips to the long, dense fur on the upperparts, which gives a silvery appearance. Despite this, the upperparts usually contrast strongly with the white underparts, although some individuals are tawny-brown below. Often has a yellowish patch of skin around the ears, which are short and broad; their lower edge extends in a fold below the line of the mouth. The dark face and membranes are similar to those of Serotine (*page 146*) and Northern Bat, although these species do not have a silvery back.

Flight: Rapid and direct on narrow, pointed wings, rather like a Noctule (*page 148*). Tends to fly at 10–40 m above ground level.
Sounds: Slow, low-frequency calls, similar

to Serotine (*page 146*) and Leisler's Bat (*page 152*), with strong constant-frequency 'tail' at 24–26 kHz.

Signs: Summer roosts: Buildings.
Winter roosts: Buildings, often tall (*e.g.* tower-blocks, churches).

Habitat: Various, including built-up areas and open fields; sometimes over water.

Habits: Hibernates from November to March. Nocturnal, emerges late dusk. The males display in autumn around tall buildings. Males and females occupy different hunting grounds.

Population and status: A rare migrant, with about 25 records and six more from offshore oil-rigs.

Key morphological details

FOREARM: 40·8–50·3 mm. TRAGUS: short and widened towards the tip. PENIS: long and thin (Northern Bat: broadly oval). Nipples: 4 (all other British/Irish bats have 2).

Kuhl's Pipistrelle

Pipistrellus kuhlii

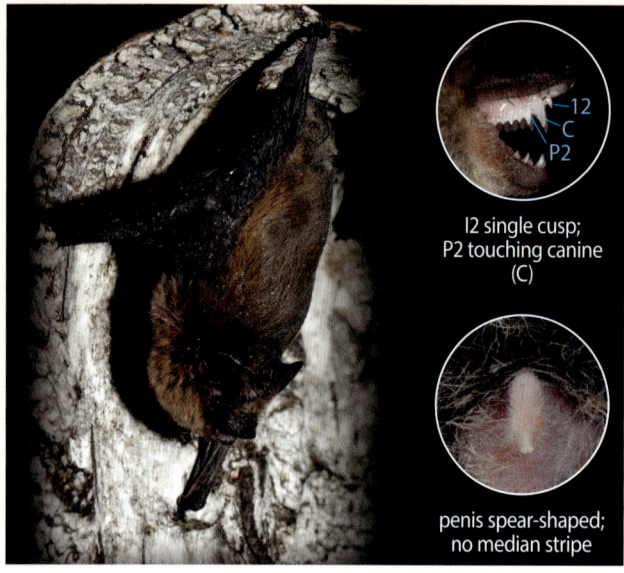

I2 single cusp;
P2 touching canine
(C)

penis spear-shaped;
no median stripe

LEGALLY PROTECTED (UK)	
Vagrant	
HB:	4·0–4·7 cm
WS:	21–26 cm
Wt:	10 g

This bat has been spreading northwards from the Mediterranean since the 1980s and now breeds in the Channel Islands. It is still very rare in Britain, but further records (or even colonization) may be expected if the range extension continues.

DENTITION
First incisor (I2) has single cusp; P2 in contact with C

Identification: Slightly larger than Common Pipistrelle (*page 154*), usually paler brown above and creamy-white below. In addition, has paler, reddish-brown ears and face (darker when young) and a sharply defined, 1–2 mm wide, white trailing edge to the wing between the foot and the fifth finger. Dentition is diagnostic (see above and key morphological details)

Flight: Fast, agile, with rapid wing-beats and regular turns, at 2–10 m above ground.

Sounds: Like other pipistrelles: 'dry' clicks or 'wet' clapping sounds with a peak frequency of 36–40 kHz (as Nathusius' Pipistrelle (*page 160*)).

Signs: Summer roosts: Buildings (*e.g.* under roof tiles); cliffs. **Winter roosts:** Buildings and rock crevices; bat-boxes.

Habitat: Mainly human settlements, from villages to cities; feeds in gardens and parks.

Habits: Hibernates November–April. Often emerges early, when still light. Patrols around street lights, sometimes hunting in groups.

Population and status: Vagrant and, more often, an accidental import (*e.g.* as stowaways in shipping containers), with more than ten records; records shown on the map are believed to be genuine vagrants (solid circles) or possible vagrants (open circles), as there is some doubt as to their origin.

Key morphological details

FOREARM: 30·3–37·1 mm. WING VENATION: as Common Pipistrelle (see *page 154*). EAR: rear margin with obvious indentation. DENTITION: on upper jaw **single cusp on first upper incisor** (I2) (all other pipistrelles have two cusps) and the canine and large second premolar (P2) are in contact – obscuring first premolar (P1) when viewed from outside (a feature shared with Savi's Pipistrelle). PENIS: **spear-shaped and without median stripe** (unlike Common Pipistrelle).

Savi's Pipistrelle

Hypsugo savii

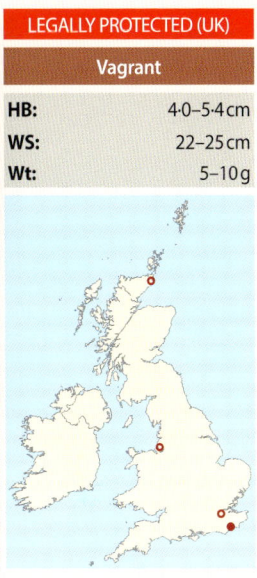

LEGALLY PROTECTED (UK)	
Vagrant	
HB:	4·0–5·4 cm
WS:	22–25 cm
Wt:	5–10 g

A Mediterranean species that has been recorded in Britain at least once in a natural state: a female found exhausted in East Sussex in January 1993.

Identification: A very small bat with longer, 'fluffier' fur than *Pipistrellus* species (*pages 154–161*) and contrastingly dark upperparts and paler underparts. The fur on the back is dark brown at the base, but the tips vary in colour from a typical golden-brown to dull mid-brown. The ears are wider and rounder than those of other pipistrelles, and the tragus much broader, while the face, ears and wings are blackish and shiny. In the hand, identified by dentition and long tail, and males by the unique shape of the penis (see key morphological details). Particoloured (*page 187*) and Northern (*page 186*) Bats look similar but are larger.

Flight: Tends to fly faster, straighter and generally higher than other pipistrelles, often above houses and over treetops, with fewer sharp changes in direction.

Sounds: Peak frequency is 32–37 kHz, lower than most pipistrelles, although just within the range of Nathusius' Pipistrelle (*page 160*).

Signs: Summer/Winter roosts: Crevices in rocks, cliffs and walls.

Habitat: Varied, including rocky areas, coasts, towns and cities.

Habits: Hibernates in winter. When active, emerges just after sunset, occasionally earlier. Migratory movements unknown.

Population and status: Vagrant, with at least one reliable natural record and others of less clear provenance (open circles on map).

Key morphological details

FOREARM: 31·4–37·9 mm. TRAGUS: short, broader than on Common Pipistrelle (*page 154*).
DENTITION: The first upper canine and large (second) upper premolar are in contact, obscuring the first upper premolar when seen from the outside. Among the pipistrelles, this feature Is shared only with Kuhl's Pipistrelle (see *opposite*), although that species has a single cusp on the first incisor, not two.
PENIS: bent at a right angle midway down (a diagnostic feature), and has long hairs.
TAIL: last vertebra extends 4–5 mm beyond tail membrane.

European Free-tailed Bat

Tadarida teniotis

LEGALLY PROTECTED (UK)	
Vagrant	
HB:	8·1–9·2 cm
WS:	25–50 cm
Wt:	25–50 g

A fairly common Mediterranean bat that specializes in catching moths high up in the open air. It has been recorded twice in Britain: one at Helston, Cornwall in March 2003 and one on the Isle of Wight in September 2016.

Identification: Uniquely (in Europe), the tail projects well beyond the tail membrane. Very large, with forward-facing, egg-shaped ears which meet in the middle, and relatively large eyes. The fur is soft and velvety, greyish-brown in colour and paler on the underparts.

Flight: High, fast and direct, 10–300 m above ground, often in wide circles.

Sounds: Echolocation calls are audible to many people, with a peak frequency at 10–14 kHz.

Signs: Colonies in Europe have a distinctive smell, like herbs. **Summer/Winter roosts:** Usually in rock crevices; sometimes caves and cracks in bridges and tall buildings.

Habitat: Typically in rocky areas, but will forage high over any habitat, including urban, areas and coasts.

Habits: Emerges in the early dusk, when it is able to hunt by sight. Seems not to be migratory, and barely hibernates – in the north of its normal range (Switzerland) torpor lasts for only up to eight days.

Population and status: Vagrant: two records.

Key morphological details

FOREARM: 57·2–64·1 mm. EAR: long and broad, the two meet at front at their base. TRAGUS: very short, but the rear margin of the ear shows a prominent rectangular flap of skin (anti-tragus). MUZZLE: long, with five folds on either side. TAIL: **long, over a third extending beyond tail membrane.**

Hoary Bat

Lasiurus cinereus

LEGALLY PROTECTED (UK)	
Vagrant	
HB:	10–14 cm
WS:	40 cm
Wt:	20–38 g

This American bat is on the British list from a specimen picked up on South Ronaldsay, Orkney, in September 1847. It is a long-distance migrant and has also been recorded in Iceland, so could again reach Britain or Ireland unaided.

Identification: A large bat with thick, 'frosty'-tipped fur, hence its name: its grizzled appearance, with an obvious yellowish collar, is very distinctive. The fur extends to cover the tail, and the ears are thickened, short and rounded with black, naked rims. Particoloured Bat (*page 187*) is somewhat similar, but smaller, lacks the yellowish collar and is white underneath.

Flight: Fast, powerful flight on long, narrow wings. It often flies very high during migration.

Sounds: Flat, low-frequency calls with a peak frequency of 20·1 kHz.

Signs: Summer/Winter roosts: Among foliage, or on bark in winter.

Habitat: Woodland, including coniferous forest. It forages over open areas in woodland, including ponds, and sometimes hunts around street lights.

Habits: Emerges late in the evening. Strongly migratory in North America, wintering in the south-west and migrating across much of the continent, females preceding the males by a month.

Population and status: Vagrant: one record.

Key morphological details

FOREARM: 50·7–56·8 mm. EAR: rounded, <25 mm long; haired inside. TRAGUS: short and broad.

Geoffroy's Bat

Myotis emarginatus

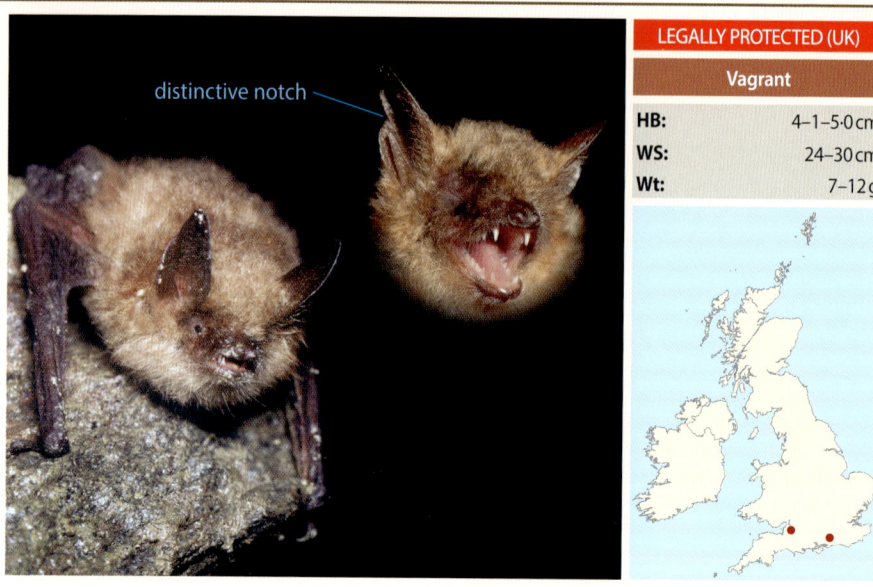

distinctive notch

LEGALLY PROTECTED (UK)	
Vagrant	
HB:	4·1–5·0 cm
WS:	24–30 cm
Wt:	7–12 g

The newest addition to the British terrestrial mammal list, the first record of this bat came as recently as 2012 with the discovery of one in Sussex (which returned in 2013) and one in Somerset, also in 2013, leading to speculation that the species may be a rare resident.

Identification: A medium-sized bat with a very distinctive right-angled fold or notch on the outer edge of its ears. Otherwise similar to Natterer's Bat (*page 172*) but has a straight calcar and shorter tragus, and lacks the sharp contrast between the upperparts and underparts – the upperparts being more reddish-brown and the underparts, yellowish-brown, rather than white or buff.

Flight: Agile and highly manoeuvrable, although not especially fast. Typically 1–5 m above ground or water.

Sounds: Frequency-modulated calls fall from over 140 kHz to about 38 kHz in 1·5–4·0 ms, with peak frequency at 55–68 kHz.

Signs: Summer roosts: Roofs of buildings (caves in southern Europe). **Winter roosts:** Underground in caves and tunnels.

Habitat: Deciduous forests, also uses meadows and gardens with fruit trees. In Europe, often found around cowsheds.

Habits: Hibernates from late autumn until April or May. Emerges well after sunset. Mainly sedentary, with usually less than 40 km (maximum 105 km) between roosts.

Population and status: Vagrant, or perhaps a new or overlooked resident, with two confirmed records, one of which involved an individual returning to the same site in consecutive years.

Key morphological details

FOREARM: 36·1–44·7 mm (usually >37 mm). TRAGUS: pointed; just over half the length of the ear (two thirds in Natterer's Bat), not reaching notch on outer edge of ear. CALCAR: straight (not S-shaped as in Natterer's Bat). TAIL MEMBRANE: short, soft hairs along edge (not as long nor as copious as in Natterer's Bat).

NT **Pond Bat**   *Myotis dasycneme*

LEGALLY PROTECTED (UK)	
Vagrant	
HB:	5·7–6·7 cm
WS:	20–30 cm
Wt:	14–20 g

This mainland European species has been recorded once in Britain, in Ramsgate, Kent in September 2004. Like Daubenton's Bat, it takes much of its food from the water surface. It has the largest feet of any European bat.

Identification: Resembles a big Daubenton's Bat (*page 174*), with similarly large feet, dense fur, relatively short, broad ears and short tragus. However, it has greyer-brown fur without any reddish tinge, more sharply demarcated whiter fur on the underparts and a reddish-brown face (darker in juveniles, which also have a dark spot on their lower lip, as in Daubenton's Bat). In the hand, told from Daubenton's Bat by its longer forearm and shorter tragus (see key morphological details).

Flight: Tends to fly a little higher and faster than Daubenton's Bat, following a straighter course.

Sounds: Frequency-modulated calls drop from 65–85 kHz down to 25–35 kHz in 4–8 ms, with peak frequency at 35–45 kHz. Indistinguishable from Daubenton's Bat.

Signs: Summer roosts: Buildings. **Winter roosts:** Underground in caves, cellars and tunnels.

Habitat: Forages over wetlands, especially those relatively clear of vegetation, including marshes, lakes and canals.

Habits: Hibernates from October to mid-March. A medium-range migrant, moving up to 300 km between summer and winter roosts.

Population and status: Vagrant: one British record.

Key morphological details

FOREARM: >43 mm (≤42 mm in Daubenton's Bat). TRAGUS: very short and slightly bent inwards, the tip being rounded; it does not reach half the height of the ear as it does in Daubenton's Bat. FOOT: large, longer than half the length of the tibia, with long bristles.

Bats that have reached Britain by assisted means

Four species of bat have been recorded in Britain, having almost certainly arrived in shipping containers. Since most of these bats could potentially be confused with native species, were not deliberately introduced, and could potentially be recorded again, they are included in this book for completeness.

Big Brown Bat

Eptesicus fuscus

Accidental import	
HB:	6·0–10·2 cm
WS:	22–27 cm
Wt:	7–13 g

Morphological details:
FOREARM: 35–42 mm.
TRAGUS rounded and short
(7–8 mm). CALCAR: lacks keel.

Sounds: Echolocation
a frequency-modulated
sweep from 80–40 kHz; peak
frequency at 45 kHz.

A very common bat of woodland, farmland and urban areas in America which has been recorded twice in Britain, at docks (in 1996 and 2000). It is rather sedentary, and not a good candidate for natural vagrancy.

Identification: A large bat, similar to Serotine (*page 146*), but slightly smaller with shorter, broader tragus. It is dark glossy brown with dark wings, tail and ears; no hairs on tail. Flight steady, direct and fast (up to 65 km/h), at a height of 6–10 m).

Mexican Free-tailed Bat

Tadarida brasiliensis

Accidental import	
HB:	8·5–10·9 cm
WS:	29–35 cm
Wt:	10–15 g

Morphological details:
FOREARM: 35–42 mm. TRAGUS:
rounded and short (7–8 mm).
CALCAR: lacks keel.

Sounds: Echolocation
a frequency-modulated
sweep from 80–40 kHz; peak
frequency at 45 kHz.

Native to southern North America, where forms enormous roosts (up to 20 million). Although migratory, British records are clearly imports: recorded twice at docks (in Kent in August 1998 and March 2003).

Identification: Dull brown, with long tail projecting well beyond the trailing edge of membrane. This feature separates it from all British and Irish species except the extremely rare European Free-tailed Bat (*page 190*) (recorded twice), which is larger, with ears meeting in the middle. Flies high and fast on long, narrow wings with pointed tips.

[NOT ILLUSTRATED]

Silver-haired Bat

Lasionycteris noctivagans

Accidental import	
HB:	9·8–11·7 cm
WS:	27–31 cm
Wt:	9–12 g

Morphological details:
FOREARM: 37–44 mm.

Sounds: Echolocation
signals at 25–35 kHz.

Scarce but widespread in deciduous and coniferous forest in the USA; has been recorded three times in Britain as an accidental import, once in a transport plane. Although migratory, not a likely candidate for natural vagrancy to Britain or Ireland.

Identification: A dark brown, almost black, medium-sized bat with frosted tips to the fur on the upperparts. The fur extends well onto base of tail membrane, distinguishing it from Big Brown Bat. The ears are short and rounded. Flight slow, at heights of up to 7 m, often flying repeatedly over the same circuit.

Little Brown Myotis (Little Brown Bat)

Myotis lucifugus

Accidental import	
HB:	6·0–10·2 cm
WS:	22–27 cm
Wt:	7–13 g

Morphological details:
FOREARM: 35–42 mm.
TRAGUS: rounded and short (7–8 mm). CALCAR: lacks keel.

Sounds: Echolocation a frequency-modulated sweep from 80–40 kHz; peak frequency at 45 kHz.

A very common North American bat that is closely associated with water. It has been recorded once in Britain, at Ipswich Docks, Suffolk, in September 1992. Although migratory, it is unlikely to be able to reach Britain or Ireland naturally.

Identification: A small bat with long, glossy, dark fur, and dark brown face, ears and wing membranes. Long hairs on hind feet project beyond the claws. Extremely similar to Daubenton's Bat (*page 174*), feeding in a similar manner by scooping up prey in wing and tail membranes. Flight fast and wavering, sometimes following a circular route.

195

Terrestrial carnivores

These pages provide an at-a-glance comparison of the small and medium-sized carnivores that occur in Britain and Ireland. Several are quite similar to each other and often give only brief glimpses as they disappear into vegetation or into the water. The seals and Walrus are also carnivores, but are covered in this book in the marine mammals section (*pages 270–279*).

CATS *pages 198–201*

DOGS (Fox) *pages 202–205*

MUSTELIDS (Otters, Weasels and Badgers) *pages 206–225*

BADGER

PINE MARTEN

POLECAT

STOAT

WEASEL

AMERICAN MINK

OTTER

See also: Tracks *p. 48*

Wildcat

Felis silvestris

Extremely shy and inhabiting remote Scottish landscapes, the Wildcat is exceptionally difficult to see: to encounter one is among the greatest privileges of British wildlife watching. However, over the years many thousands of domestic cats have reverted to living wild. These feral cats are at large everywhere, including Scotland, and frequently hybridize with their wild counterparts, resulting in a loss of genetic integrity of the wild form and often causing identification difficulties.

LEGALLY PROTECTED (UK)		
Biodiversity List (Sc)		
Very rare		
HB:		47–66 cm
T:		26–33 cm
Wt:		3·8–7·3 kg (male),
		2·4–4·7 kg (female)

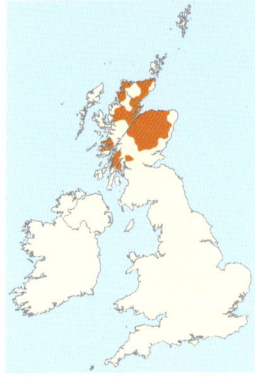

Identification: Identification as a cat is simple, given a good view: the problem is determining a true Wildcat from a domestic or Feral Cat (*page 201*), especially a tabby with similar patterning, and particularly from a hybrid between the two. A true Wildcat is generally much bigger and chunkier, although in practice this feature may be difficult to determine in the field. Other pointers towards a true Wildcat, especially when taken together, include a thick tail with a rounded, black tip; the tail encircled with regular black rings; the dorsal stripe on the lower back ending at the base of the tail (not extending down the tail as on some Feral Cats and hybrids); and unbroken rump and flank stripes. Features that unequivocally point towards hybridization or domesticity are the presence of white patches on the mouth, throat or chest; stripes breaking up into spots; and the dorsal stripe extending down onto the tail. However, it may not be possible to ascertain true provenance in some cases without genetic analysis.

WHERE TO LOOK/OBSERVATION TIPS

One of the hardest land mammals to see in Britain. Considerable persistence is required, and you will need to be out at night, dawn or dusk driving remote roads or hiking in rough country. Best to try forest edges and moorland fringes, although sometimes hunts on scree slopes. The Ardnamurchan Peninsula has long been the traditional site for those hoping to encounter this enigmatic species.

Sounds: Mews and purrs like a domestic cat, but rarely heard in the wild.

Signs: Main signs are tracks, droppings and scratch marks; prey remains are also identifiable. The eyeshine is pale green. **Tracks:** Although there are five toes on the front paws and four on the hind paws, the inner toe on the front foot does not leave a print. The fore and hind prints are therefore very similar, 4–6 cm long × 3·5–5·0 cm wide, with a roughly triangular, three-lobed interdigital pad and four digital pads radiating out in a circle; the claws are retracted and do not show. Stride 30–60 cm. Domestic cat

Wildcats raise their litters in a den, often among rocks or tree roots.

prints are smaller (3·5–4·0 cm long × 3·5–4·0 cm wide), with a relatively smaller interdigital pad and more distant digital pads. **Droppings:** The scat is cylindrical, 4–8 cm long × 1·5 cm wide, with pointed, twisted ends. Often segmented and containing bone, hair and insect remains, it has a pungent, musty smell. Sometimes left in the open on rocks and other prominent places, but may be covered up with vegetation. **Scratch marks:** Wildcats scratch their claws on trees or posts (as do domestic cats), leaving distinctive vertical scars low down. **Prey remains:** A selective feeder, often eating just flesh, leaving unbroken bones behind; and may just tear off the head and eat the brain. Smaller prey may be partly buried.

Habitat: Occupies the transition zone between forest and open country such as moorland and mountain slopes. Usually hunts in woodland clearings, along forest edges, in young plantations and on open moorland, rather than deep within forests; open and high areas are used more often in summer.

Food: Mainly small rodents, Rabbits and hares, but also birds, amphibians and fish. Often waits outside rodent holes for the occupant to appear, whereupon it lunges and bites the back of the neck. Otherwise stalks prey, hiding in cover before a final sprint. Eats grass to prevent hair balls.

Habits: Active throughout the year, principally during the night and twilight, although can be much more diurnal in winter. Snowfall, rain and high winds impede hunting. Mainly solitary, with both sexes holding territories exclusive from their congeners; males hold a territory that overlaps those of several females. The size of territory is prey-dependent, up to 14 square kilometres in moorland to less than 1 square kilometre in Rabbit-rich areas. Wildcats mark their territories by spraying urine on trees and rocks, and leaving prominent piles of droppings. Occupies dens among boulders or tree roots, or in old Fox earths or Badger setts; dens usually provisioned with little bedding.

A true Wildcat is large and stocky, has a thick tail with a rounded, black tip and regular black rings; the black line down the top of the lower back ends at the base of the tail. In addition, there are unbroken dark stripes down the flanks and there is no hint of white on the muzzle.

Breeding behaviour: Wildcats mate between February and April, with a peak of births in May (range April–September); after mating the male takes no further part in raising the young. Usually just one litter a year, but may have a second if the first is lost. Typical litter size is 2–6, born in a den lined with grass or Heather. The kittens' eyes (blue at first) open at 7–13 days, and the youngsters can move between dens at 10–12 weeks. They are fed exclusively on milk for 6–7 weeks, with a combination of solid food and milk for some time thereafter. Having stayed with their mother for about five months, siblings generally disperse at the same time and go their own way.

Population and status: Found only in Scotland, the number of true Wildcats is uncertain, estimates ranging from fewer than 400, up to 1,000. Once widespread in Britain, they have been persecuted for centuries (and are probably still shot and trapped illegally), and frequently become road casualties. But the greatest threat now is hybridization with feral cats, such that few, if any, genetically 'pure' Wildcats remain away from isolated havens, such as the Ardnamurchan and Morvern peninsulas.

Wildcat and Feral Cat compared

		Wildcat	Hybrid	Feral Cat
BACK STRIPE		Stops at tail base	Runs onto tail	To tip of tail (or absent)
TAIL RINGS		Distinct, contrasting	Indistinct, merged	Merged with back stripe (or absent)
TAIL TIP		Blunt and thick	Intermediate	Tapered and thin
FLANK	STRIPES	less than 25% broken	25–50% broken	over 50% broken (or absent)
	SPOTS	None	Some	Many (or absent)
NAPE		4 thick stripes	Intermediate	Thin stripes (or absent)
SHOULDER		2 thick stripes	Intermediate	Indistinct stripes or none

Hybrid Wildcat × Feral Cat; although this individual shows many of the characteristics of Wildcat, its build and proportions are relatively slim.

Feral Cat

Felis catus

Many domestic cats live, at least partially, independent of their owners, and a large feral population has become established. Individuals vary considerably in size and coloration, but in some cases, the colour and pattern can be almost identical to that of the native Wildcat. The two species may be indistinguishable in parts of northern Scotland where their distribution overlaps; here, hybridization is frequent.

Very common	
HB:	43–56 cm
T:	23–34 cm
Wt:	3·5–7·0 kg (male),
	2·5–4·5 kg (female)

Kept as a pet throughout, but free-living populations are established, particularly in urban areas (*e.g.* London, South Wales, Greater Manchester, Central Lowlands of Scotland). Some truly feral populations on islands (*e.g.* Outer Hebrides).

Identification: Readily identified as a cat, but in northern Scotland a feral 'tabby' cat could be easily confused with a Wildcat (*page 198*). However, Feral Cats and hybrids, are typically smaller and of a lighter 'build' than Wildcat, have a more pointed tail, and rarely the same pattern of stripes; they usually show some white patches.

Sounds: Mews and purrs.

Signs: Tracks, droppings, scratch marks; identifiable prey remains. Eyeshine is pale green. **Tracks:** As for Wildcat but smaller at 3·5–4·0 cm long × 3·5–4·0 cm wide and with a relatively small interdigital pad and more distant digital pads. **Droppings:** Tend to be buried if near human habitation, but elsewhere deposited on top of conspicuous objects. **Scratch marks:** Distinctive vertical scars low down on trees or posts. **Prey remains:** Like those of Wildcat.

Habitat: Wide variety: urban brownfield sites, woodlands, agricultural areas, *etc.*

Food: Scavenges. Also takes rodents, Rabbits, hares and birds.

Habits: Active year-round, mainly at night, but can be diurnal in winter. Mainly solitary; both sexes territorial, those of males overlapping with those of several females. Lives in dens, as Wildcat.

Breeding behaviour: Mates mainly during the early spring, with a peak of births in May; females rear the young alone. The litter size is 2–6 and the young are born in a den.

Population and status: Probably very numerous but poorly known; some estimates suggest that there are up to 1·5 million in the UK.

Tabby cats can look very like Wildcats, although they are smaller and more delicately proportioned.

WHERE TO LOOK/ OBSERVATION TIPS

Urban developments, dockyards, farm outbuildings, sheltered rocky or wooded areas.

See also: Droppings *p. 42* | Tracks *p. 48*

Fox

Vulpes vulpes

(Red Fox)

One of our best-known animals, the Fox is found throughout much of the Northern Hemisphere, everywhere from treeless tundra to semi-desert, reflecting its adaptability and opportunistic nature; in Britain and Ireland it is still common and widespread, despite years of persecution.

Very common	
HB:	58–90 cm
T:	32–49 cm
Wt:	3–11 kg
Males slightly larger and heavier than females	

Identification: Resembles a medium-sized, low-slung dog, but has a long, bushy tail, a slender muzzle, pointed ears and usually a rich russet body colour, often mixed with grey or yellowish. The tail has a white tip, and the throat and lower muzzle are pure or dingy white, while the feet and back of the ears are blackish. The fur is thicker in winter. Moves with a brisk but relaxed trot, holding its body horizontal and the tail above the ground. Familiar, and unlikely to be confused with any other British wild animal.

Sounds: Has a wide vocabulary but two sounds are particularly well known: a plaintive and rather melancholy volley of three barks, and a loud scream uttered by the vixen in winter. The scream has a yelping quality, and has been mistaken for that of a human in distress.

Signs: Signs include tracks, droppings, dens and feeding signs. Eyeshine is typically light orange, but may appear white head-on. **Tracks:** Prints 5–7 cm long × 4·0–4·5 cm wide; four toes show on each foot. Recognisable by the oval shape of the print; the semicircular interdigital pad being slightly smaller than the digital pads; and the two central toes

The white-tipped tail is diagnostic.

WHERE TO LOOK/OBSERVATION TIPS

Easiest to see in towns and cities, where it is used to people, and routinely visits gardens; can be very shy in rural areas. Often seen if you drive through a suburb or city centre at night.

which extend ahead of the others. The stride is about 45 cm, and tracks are often in a straight line. In contrast, the track of a domestic dog usually shows a wider pattern of prints, while the individual prints are more rounded, with digital pads splayed around a very large interdigital pad. **Droppings:** Pointed, twisted and segmented, with hairs holding the segments together. Usually dark-coloured, scats are 5–20 cm long × 1·5–2·0 cm wide, and have a strong musty smell when fresh. Typically they contain some recognisable indigestible remains such as hair, feathers or fruit stones. Placed in prominent locations such as upon rocks, molehills or in the middle of a path. Domestic dog droppings vary greatly but are typically more sausage-shaped, untwisted, lack the musty smell and do not show coherent remains of what has been eaten. **Dens (or 'earths'):** Excavates a single burrow 25–30 cm wide, or takes over Rabbit or Badger burrows, as its den. Will also use rock crevices and other natural holes, and in gardens the gaps under sheds and other buildings. The vicinity of a Fox den lacks the well-worn paths leading to a Badger sett, and prey remains are often left outside the burrow. Active dens always have a distinctive, pungent odour. **Feeding signs:** Bird carcasses from Fox kills are distinctive: the wing feathers are cut near the base, whereas a bird of prey leaves them untouched. Eggs eaten by a Fox have one end bitten off and are often stored away under vegetation. Mammal carcasses have broken bones, and the skins may be partially turned inside out; they are often at least partially buried. The heads of some animals may be bitten off.

Habitat: Occurs in almost any habitat, but especially within mixed landscapes. Found in woodland of all kinds, scrub, farmland, heathland and moorland, dunes, coast and marshes, and sometimes abundantly in built-up areas, most notably suburbs with larger houses and gardens.

In summer the coat looks sleeker than in winter.

Food: Although predatory, its mixed, opportunistic diet also includes carrion (sometimes foraged from dustbins), fruit and berries. Prey includes mammals such as Rabbits, hares, rats, mice and voles, and birds (particularly ground-nesters), as well as significant numbers of invertebrates, especially worms and beetles, often located by ear: individual animals often have their own feeding preferences and specializations. Small mammals, detected by sound, are captured in the front paws after a spring and pounce, and killed with a bite. Foxes often cache excess food – including whole carcasses – for later consumption, using many different hiding places. When faced with concentrations of potential food (for example in seabird colonies or hen-houses), individuals may go on a killing spree.

Habits: Active year-round; usually nocturnal, with peaks of activity at dawn and dusk. Foxes usually live in a joint territory based on a family unit: a male and female with their most recent young, some of which may remain and help with guarding, grooming and feeding the cubs in the following breeding season. The group territory size varies, typically 30 ha in urban areas and 270 ha in more rural places, but sometimes much more. Territories are marked with urine and droppings; when urinating, dog Foxes cock a leg, vixens usually squat. There are several dens in each territory but they are not used all the time, the animals often simply lying up in dense vegetation by day. May also bask in the open where undisturbed. Fights over territory do take place, but this is usually prevented using signals such as gestures and body posture.

Breeding behaviour: Mating occurs between December and February; a female is on heat for three weeks, during which time she is usually closely accompanied by a male. Copulation is repeated and prolonged: indeed the male may become 'locked-in', unable to withdraw for more than an hour. Males often visit the territories of Foxes nearby, where they will mate with other females. Litters comprise 4–5 cubs, which are usually born in March (range from February to April); the cubs' eyes open at 11–14 days, they are suckled for four weeks and start to catch their own food by 5–6 weeks of age. The cubs become progressively independent and some leave the territory within their first year.

Population and status: Common and widespread, and absent only from some islands, the British population is stable at about 258,000, and there are at least 150,000 in Ireland. The Fox is not protected and many are shot or poisoned; it was once widely hunted with dogs but this practice has been illegal in the UK since 2004. Foxes killed on roads are a frequent sight.

When hunting small mammals, Foxes habitually pounce from a distance.

▲ Fox cubs regularly engage in play-fighting, often near the entrance to the den.
▼ The area around the den is often flattened by the activity of the cubs.

Badger

Meles meles

(European Badger)

Britain's damp and mild climate suits the Badger, and parts of western Britain have the highest density of these animals in the world, a reflection of the mixed landscape in which they live and the abundance of their main food source, earthworms. Despite being persecuted for centuries, the Badger is still common throughout Britain and Ireland. Although now legally protected, its survival owes much to its secretive nature: it is wary and predominantly nocturnal, and spends much of its time underground in sometimes extensive burrow systems, called setts.

LEGALLY PROTECTED	
Common	
HB:	56–89 cm
T:	11–20 cm
Wt:	8–20 kg

Males slightly larger and heavier than females

Identification: Britain's largest terrestrial carnivore; unmistakable if seen well. Has a black-and-white-striped head, silver-grey body fur and a thickset appearance. The feet are black, the tail pale and the ears have white tips. The overall colour of the body fur can vary, with some individuals having a browner appearance, especially in summer. Various colour forms also occur, including albino (white), melanistic (black) and reddish (erythristic). It has very short legs and moves with a distinctive, unhurried trot; it can also climb to a limited extent.

Sounds: Quite noisy, especially at night, producing growls and whickering sounds.

Signs: Leaves many signs, including tracks and trails, droppings, large dens (setts), feeding signs, scratch marks on logs or tree trunks, and hair caught on wire fences. Eyeshine is green, very low to the ground. **Tracks:** Prints of fore feet 6 cm long × 5·5 cm wide; hind feet 6 cm × 5 cm. Often well formed in mud, they differ from all similar tracks in that all five digital pads and long claw marks lie in an arc in front of the large, kidney-shaped interdigital pad, like a small human hand. Worn paths leading to and from the sett are often obvious. Badgers also make paths across pastures which, unlike deer or human paths, do not deviate when they reach a fence. **Droppings:** Elongate, sausage-shaped, up to 10 cm long × 2 cm wide, but unlike those of most carnivores not strongly twisted or coiled. Often rather loose and muddy after earthworms have been consumed; incorporated remains include hard insect parts, bones and fur. Droppings are deposited in characteristic latrines – pits 10–15 cm deep, and not covered; often on territorial boundaries. **Sett:** The sett consists of a series of large holes, typically dug into a slope in or along the edge of woodland, although may be excavated in flat ground where there is little risk of flooding. The holes of a main sett are interconnected below ground, and may extend over a distance of 100 m, with up to fifty entrances. Satellite setts are smaller, with fewer entrances, and used less frequently. Setts are marked by spoil heaps of excavated soil, usually mixed with bedding

WHERE TO LOOK/OBSERVATION TIPS

Most often seen casually during drives in the countryside at night. Otherwise try visiting advertised Badger-watching hides or a garden in which Badgers are fed. Less reliable, but more enthralling, is to wait outside an active sett. You should arrive about an hour before dusk (ideally between March and October) and ensure that you remain still, quiet and downwind from the animals; standing on a raised platform such as a fallen branch helps, or you might climb into a nearby tree (taking due care).

▲ The Badger is instantly recognisable, due to its black-and-white-striped face.
▼ Although usually nocturnal, can sometimes be seen during daylight.

such as straw or Bracken: fresh spoil, and lack of debris or cobwebs across the entrance, are signs of recent activity. Well-trodden paths often run from setts to latrines and regular foraging areas. Setts are most obvious in winter and early spring, before vegetation has grown up. The holes are usually more than 25 cm across, and broader than they are tall, whereas Rabbit burrows are smaller and more circular. The entrance to a Fox den is similar in size to a Badger's, but solitary, and usually has a strong musty smell. Both Foxes and Rabbits may use the less frequented parts of a Badger sett. **Feeding signs:** On lawns and pasture, Badgers leave conical 'snuffle holes' where they have dug to find worms. May also strip Sycamore bark for sap. Damaged or excavated bees' or wasps' nests are also often the result of Badger feeding.

Habitat: Requires a combination of deciduous woodland and nearby pasture, the latter used for foraging for earthworms. Also found in gardens, even in urban areas. Badgers avoid places where the ground is particularly wet, such as marshes, and are generally less numerous in upland areas (often absent).

Food: A broad diet, albeit based on earthworms, which typically make up 50% of the bulk. Larger insects, such as beetles, leatherjackets, caterpillars, bees and wasps are the next most important item; other meat includes small mammals (voles, mice), Hedgehogs, ground-nesting birds and carrion. Other foodstuffs include cereals, fruit, berries and nuts (including peanuts, which people often use to tempt Badgers into gardens).

Habits: Does not hibernate, but spends much time underground from November to February, emerging only on mild nights. Outside the winter period, typically emerges around dusk, grooming for a while before departing on a foraging trip. Easily disturbed by human activity at the sett, retreating underground, but in places with little disturbance often adopts a more diurnal activity pattern. The sett is excavated using the powerful feet and claws, and can occupy several levels, sometimes with more than 300 m of tunnels. The sleeping and breeding chambers are lined with bedding, such as Bracken and dried grass. This is often collected some distance away and the Badger will drag a bundle of bedding backwards to the sett with its forepaws. Each sett is occupied by a social group of adults of both sexes, on average six, but sometimes many more. The group occupies a territory of 30–150 ha, delineated by latrines and well-trodden pathways, and scent-marked with dung, urine and glandular secretions. Males, especially, will fight to keep members of other social groups out of their territory.

Breeding behaviour: Mating usually takes within the social group, and in spring, but can occur in any month. Implantation of the eggs is delayed until December, with most cubs born in February (range January–March). There are up to five cubs in a litter; their eyes open at five weeks, they first venture out from the burrow at eight weeks and are weaned at 12 weeks. Throughout this time they are frequently groomed and looked after by all members of the social group. Young usually remain with their social group, but some dispersal can take place.

Population and status: Currently about 300,000 in Britain, having been increasing for several decades. Badgers are often killed on roads, and, although they are protected, some illegal persecution takes place, including the ongoing practice of Badger-baiting (a 'sport' in which dogs draw a badger from its sett and kill it), even though this has been illegal in the UK since 1830. In Britain, a government-funded campaign of localized culling has been implemented in a controversial attempt to control Bovine Tuberculosis. In the Republic of Ireland, where the population is estimated at 84,000, there has been an ongoing culling programme since the 1980s. However, this has now officially been stopped and a vaccination programme has begun instead.

▲ Badgers are low-slung animals and readily wear paths through vegetation.
▼ Entrances to Badger setts are wider than they are high.

 **Otter**

Lutra lutra

(European Otter)

In the 1970s, the Otter population was at a very low ebb: it was absent from much of England and had declined drastically in parts of Scotland and Wales. Poor water quality, in particular pollution by organic pesticides and polychlorinated biphenyls, which accumulate up the food chain, is believed to have been the main contributory factor. Subsequent controls on water pollution and legal protection against persecution have enabled the population to recover, aided by reintroduction schemes using captive-bred animals; it is now thriving over most of its former range. The population in Ireland has, however, remained fairly constant.

LEGALLY PROTECTED	
Biodiversity List (En, Wa, Sc, NI)	
Locally common	
HB:	60–90 cm (male), 59–70 cm (female)
T:	35–47 cm
Wt:	5–17 kg

Identification: A relatively large, long-bodied aquatic animal with a distinctive broad-based, tapering tail. It has a dog-like face, dense bunches of whiskers and a uniformly brown body except for a pale throat and upper belly. The fur may look smooth and shiny or matted into clumps. Lively and apparently playful, Otters often roll around together on land, and slide on wet mud and vegetation. When diving in still water, it usually rolls smoothly forward with a pronounced flick of the tail as it disappears, but often dives with a forward leap in more turbulent conditions. Much larger than American Mink (*page 214*), Polecat (*page 218*) and Stoat (*page 220*); compared with the semi-aquatic American Mink, it swims much lower in the water, often with just the head showing, whereas the mink appears more buoyant, holding its head and back out of the water. When swimming quickly, an Otter may adopt a muscular 'porpoising' action, leaving a substantial wake. On land it progresses by walking and running, as well as with the bounding gait typical of the smaller mustelids.

Sounds: A loud whistle that can be clearly heard above the sound of flowing water; used in maintaining contact between the female and her young, and between the male and female during courtship. At close quarters, an anxious 'huff' may be heard.

Signs: Include droppings (and their effect on vegetation), tracks, slides and feeding signs. The eyeshine is dull red. **Droppings:** Known as spraints: usually black, with a tarry texture when fresh, smelling sweet and musky (reminiscent of Jasmine tea) or fishy; becoming grey and crumbly (ash-like) with age. Placed on prominent sites near water, including boulders and bridge supports; also high up beneath eroded overhanging banks, and outside the dens (holts). In maritime environments with relatively poor soils, spraints can fertilize the vegetation, making it much more verdant and lush. Often contain fish bones and scales or shells. Size varies, 3–10 cm long and up to 1 cm wide, but lacks the stringy end of American Mink scats, which also have a more distasteful smell. **Tracks:** Larger than those of other mustelids, reflecting the adaptation of the feet for swimming: fore prints 6·5–7·0 cm long × 6·0 cm wide; hind feet usually longer 6·0–8·5 cm × 6·0 cm. Five digital pads (although often only four show) radiating from a semicircular interdigital; rather round overall (American Mink prints are pear-shaped), with teardrop-shaped toes and short claws. A clean, fresh print may show the characteristic webbing between the toes, and the tail may leave an impression as a broad line. Stride 40–80 cm. **Slides:** Often slide down steep banks into the water on their bellies, creating an obvious flattened track with regular use, usually linked to well-

▲ The long, tapered tail and large size distintingishes the Otter from American Mink, Pine Marten and Polecat, which have bushy tails.
▼ On the coast, often eats prey while sitting in an exposed position.

worn bankside paths. Otters also slide and frolic in snow. **Feeding signs:** Favoured feeding points usually marked by piles of uneaten food remains and spraints.

Habitat: Occurs in two broad habitats: freshwater, including rivers and lakes; and coastal areas, where they are mistakenly called 'sea otters' (actually the name of another species that occurs in the north Pacific). Even coastal animals need access to freshwater, to facilitate grooming and the removal of salt from their fur. Normally lives within 100 m of water, but will move considerable distances over dry ground between wetlands.

Food: The bulk of the diet consists of fish, including sticklebacks, European Eel and Perch, and in coastal areas Butterfish and Lumpsucker. Often ambushes fish within underwater vegetation. Also takes small mammals, birds, amphibians and crustaceans, and, in Scotland, sea urchins, according to season and availability. Catches prey in and out of water, using its whiskers as touch receptors in dark or cloudy conditions.

Habits: Active all year, the daily activity varies: coastal Otters are more diurnal, whereas riverine animals tend to be active at night, particularly in the first few hours after dusk. May spend 4–6 hours in the water during each 24-hour period, alternating foraging with rest; the average foraging trip covers 4·5 km. Males are usually solitary, inhabiting a home range of about 15 km of river, with some overlap: when males meet they usually fight, often causing wounds. Females are typically solitary along rivers (average 7 km of bank), but sometimes share a communal, defended home range with other females and their cubs in lakes or along coastlines, within which each female has its own core area. Territories are marked with prominently placed spraints. There are

A female and two cubs. Otters are often found on the seashore, especially in Scotland.

WHERE TO LOOK/OBSERVATION TIPS
Easiest to see in coastal habitats in Scotland (Mull, Skye and Shetland are particularly renowned), where it is often active by day; two hours either side of high tide is often the best time. Another approach is to watch quietly from a wetland bird hide (*e.g.* Somerset Levels), especially around dusk, or to wait patiently beside a known Otter river. Good clues to an Otter's presence include waterfowl moving away quickly, gulls mobbing overhead, and a general panic amongst riverside birds.

several holts within each home range: some are above ground, usually among vegetation, where an animal may sun itself while resting; others are in tunnels or holes, both natural and artificial, including crevices between tree roots, in piles of flood debris, or within drains and culverts. In coastal areas may use old Rabbit burrows, natural rock crevices or holes within rock-armoured sea defences as holts. Rarely digs its own burrow, and bedding is not usually brought in, except to the breeding (natal) holt.

Breeding behaviour: Can breed at any time of year, although more seasonal in the north, where mating and births are in summer. The male and female engage in courtship chases, which culminate in the male biting the female on the scruff of the neck, prior to mating. The male takes no part in bringing up the young. Litters comprise 2–3 cubs, which are born in a holt; their eyes open after 30 days, they are weaned at 14 weeks, and remain with their mother for about a year.

Population and status: Widespread and now fairly common after a recent significant increase: the estimated population is 12,000 in Ireland, and 10,300 in Britain. It is, however, still affected by pollution and development, and vulnerable to road collisions.

The Otter swims very low in the water often with just the dog-like head showing. American Mink is more buoyant and swimming looks to be more of an effort.

American Mink

Neovison vison

The American Mink was widely kept in fur farms in the mid-20th century. Some inevitably escaped and it was first reported breeding in the wild in Britain in 1956. It is universally unwelcome, predating fish, Water Voles and other wildlife, and despite intensive trapping and other control methods has become firmly established in both Britain and Ireland. Mink swim well, can dive to 5–6 m and travel underwater for up to 30 m.

Introduced; fairly common	
HB:	32–47 cm (male slightly larger than female)
T:	13–23 cm
Wt:	0·8–1·8 kg (male), 0·4–0·8 kg (female)

Identification: A blackish to dark brown, long-bodied, short-legged predator with dense, lustrous fur and a long, rather fluffy tail about half the length of the body. The ears are short, largely hidden in the fur, and it has a distinctive white chin-spot. Occupies similar habitats to the Otter (*page 210*), with which it is often confused. However, the Otter is much larger (up to twice as long), with a whitish underside (including all of the chin and the sides of the neck), a squarer head and a long, tapering, non-bushy tail. Otters swim low in the water and dive smoothly with a forward roll; American Mink, on the other hand, swim high in the water and have a weaker, more 'doggy-paddling' swimming style.

Sounds: A shrieking call, rarely heard in the wild.

Signs: Detected by droppings, tracks and feeding signs.
Droppings: Cylindrical with twisted ends, and often coiled: 5–8 cm long and up to 1 cm in diameter, often dark green or brown. Differentiated from those of Polecat (*page 218*) or Stoat (*page 220*) by the frequent presence of fish bones. Typically found on the banks of rivers and on prominent rocks, and have a foul, pungent smell when fresh (Otter spraints have a sweeter smell and a looser consistency). **Tracks:** Small, similar to those of Polecat: prints of fore feet

The bushy tail immediately distinguishes the American Mink from the Otter.

WHERE TO LOOK/OBSERVATION TIPS

Not easy to see, although perhaps more reliable than Stoat or Weasel as its habitat is more constrained. Walking along a river or canal at dusk may pay dividends, especially in late summer when the young become independent.

3 cm long × 4 cm wide; hind feet 4·5 cm × 3·5 cm. Five small digital pads with long claw marks radiate in starlike fashion from a large, lobed interdigital pad, the two largest lobes in the middle. Runs with a bounding gait, stride 30–40 cm; paw prints often paired. **Feeding signs:** Eats prey in secluded sites, such as a hollow tree, leaving a pile of fish scales, bones, egg shells and other debris.

Habitat: Mainly freshwater wetlands with bankside cover, including slow-flowing rivers, lakes, marshes and canals. Also on some estuaries and coasts.

Food: Has a broad diet, including small mammals, birds, fish and crustaceans. What it eats is largely dictated by season and by local conditions with, for example, breeding birds and their nestlings and eggs being prominent in spring and summer.

Although a range of colour forms occurs, one of the key identification features is the white chin-spot.

Habits: Active all year, although less so in winter; mainly in twilight or darkness, but sometimes during the day. Defends a territory with scent-marking, aggressive posturing (such as raising the tail), and sometimes fights between males. Home range typically 1–6 km of riverbank, those of males being larger than those of females. Occupies several dens within the home range, often in hollow waterside trees, among rocks or in Rabbit burrows close to water; not normally excavated by the animal itself.

Breeding behaviour: Mating occurs from February–April, with births in April–May. Has a single litter each year of 4–7 young; the kits are born in a den lined with fur, feathers and dry vegetation. All parental care is by the female: she brings food to the den until the young are weaned at 5–6 weeks, and teaches the young how to hunt. At about 10 weeks of age they are independent.

Population and status: The British population is around 110,000, but now declining as a result of trapping to safeguard fisheries and native wildlife, and due to increasing competition following the recovery of the Otter population. It is now close to being eliminated from some areas, including the Western Isles. The American Mink is widespread in Ireland, with an estimated population of 20,000–33,000.

Swims buoyantly with much of its back showing, and with a rather 'doggy-paddle' style.

See also: Tracks *p. 50*

Pine Marten

Martes martes

A sleek carnivore, the Pine Marten is set apart from its near relatives by its ability to climb trees – it can even hunt Red Squirrels high in the branches. Once widespread, and recently confined to just Scotland and Ireland, it is now beginning to return to several parts of its former range, from Northumberland to Staffordshire, North Wales and, more recently still, the New Forest in Hampshire.

Identification: A dark, cat-sized carnivore with a pointed muzzle, large pale ears, long legs and long, bushy tail. Its thick fur is chocolate-brown, with an ochre-yellow (paler in winter) throat patch of variable size. The Polecat (*page 218*) is smaller, with shorter legs and tail, and usually white marks on the face, while American Mink (*page 214*) has shorter legs, very small ears and (usually) an entirely dark body.

Sounds: Occasionally makes a "*tok-tok-tok*" call, and will growl and squeal during the breeding season.

LEGALLY PROTECTED
Biodiversity List (En, Wa, Sc, NI)
Uncommon

HB:	48–56 cm (male), 36–45 cm (female)
T:	17–28 cm
Wt:	600–2,200 g (male), 500–1,200 g (female)

By far the most likely small carnivore to be seen in trees.

WHERE TO LOOK/OBSERVATION TIPS

Sometimes encountered casually by walking in woods or driving Scottish lanes, especially at night, but by far the easiest way to see a Pine Marten is to attend one of the regular feeding sites.

Signs: Look out for droppings and tracks. Eyeshine is electric blue. **Droppings:** Scats 8–10 cm long and 1·0–1·2 cm wide, black, typically very twisted and sometimes folded, and tapered at one end. Often found on forest trails and roads, at intervals of perhaps every 100–200 m, and around dens. They have a slightly sweet, floral smell, unlike the musty scats of Fox (*page 202*) and American Mink, and, especially in autumn, often contain fruit stones. **Tracks:** Fore feet prints 8 cm long × 6 cm wide; hind feet 8·5 cm × 6 cm. Five digital pads radiate from a semicircular interdigital, but the inner toe does not always show; claw marks are short and blunt. Sometimes found in groups of four as the animal progresses with two-footed leaps, but this is variable; otherwise the stride length is 50–80 cm.

Habitat: Largely a forest and woodland species, which despite its name is by no means confined to conifers. Prefers structurally diverse older stands of trees with many holes and cavities. Dens are in tree holes and crevices among rocks and tree roots, but in Scotland many dens are in buildings. Can inhabit open rocky areas with dense patches of scrub, particularly in western Ireland.

Food: Largely carnivorous; voles are the most important prey, but also takes mice, shrews, squirrels, Rabbits, birds and their eggs, amphibians and insects. Eats honey, berries, nuts and fungi especially in the autumn. Also takes carrion, and visits bird tables and rubbish bins for scraps.

Habits: Climbs trees and active all year; mainly nocturnal but in summer may forage after dawn and before sunset. Lives in a den, lined with dry grass and Bracken, within a home range which is defended, sometimes aggressively, against animals of the same sex, although male and female territories may overlap. Continually marks its territory with scent, urine and scats, often in prominent places.

Breeding behaviour: Adults mate in June–August after a courtship chase; copulation involves the male holding the female by biting her neck. The male plays no part in rearing the young, and can be promiscuous. The 2–6 young (kits) are born in the natal den in late March/early April; they are fed on milk for six weeks, and emerge at 7–8 weeks.

Population and status: The Scottish population is about 3,500, with fewer than 100 (although increasing) estimated in England and Wales (where a reintroduction programme is underway); 2,700 in Ireland. Within its core range it may be threatened by woodland fragmentation.

Sometimes active in daylight, especially in the spring and summer – the ochre-yellow throat patch is unmistakable.

See also: Tracks *p. 50*

Polecat

Mustela putorius

The Polecat is one of the most difficult mammals to see in Britain, by virtue of its secretive habits. Until recently, largely restricted to Wales and the Midlands as a result of persecution, it is now greatly expanding its range. The domesticated form, the Ferret (classified as subspecies *Mustela putorius furo*), is often released and can be seen almost anywhere.

Identification: A small, dark, slim and long-bodied, bushy-tailed ground predator, most easily recognised by contrasting white marks on the short, rounded face. Although these markings are sometimes faded or absent, especially in summer, leading to possible confusion with American Mink (*page 214*), that species is always uniformly coloured, whereas Polecat usually has paler brown patches, at least on the flanks. May be impossible to distinguish from Ferret without skull examination, and hybrids occur which are almost identical to true Polecats. However, Ferrets are usually more variable in colour (usually paler) and markings. They often have one or more white feet, a creamy throat patch that is usually more than 50 mm long, and scattered white guard hairs on the body, especially on the hindquarters. On the face, the dark fur does not reach the nose, and the pale cheek marks and frontal band are often extensive, not contrasting strongly with the dark face-mask.

LEGALLY PROTECTED (UK)	
Biodiversity List (En, Wa, Sc)	
Uncommon	
HB:	29–46 cm
	(male larger than female)
T:	12–14 cm
Wt:	0·5–1·5 kg (male),
	0·4–0·8 kg (female)

Distribution in Ireland and on Isle of Man relates to feral Ferrets only; in Britain, distribution shows viable populations but may include Polecat x Ferret hybrids.

Polecat is superficially similar to American Mink, but has a distinctive face pattern and a rusty-brown patch on the flanks.

Committed mammal-watchers will spend hours driving country roads at night hoping to glimpse this animal running across their path, especially in its heartland of Wales. It is sometimes possible to see it around farm buildings at dusk.

Sounds: Usually silent, but can squeal, hiss and scream.

Signs: The most reliable Polecat signs are scats, but tracks and feeding signs are recognisable. **Droppings:** Scats are cylindrical, often slightly coiled, twisted and with one end tapered to a string-like tip; up to 7 cm long and 1·5 cm across. They are usually black, smell very strongly musty, and include fur and bones. Often placed in latrines, which may be in outbuildings. Scats are larger than those of Stoat (*page 220*) or Weasel (*page 224*); similar in size to those of American Mink, but do not contain fish bones. **Tracks:** Small, similar to American Mink; prints of fore feet 3·5 cm long × 3·5 cm wide; hind

Domesticated Polecats are known as Ferrets and, although they occur in a range of colours, often closely resemble their wild ancestor.

feet 4·5 cm × 4·0 cm. Five digital pads radiate out from an almost triangular interdigital pad; the outer toes are set back close to the interdigital. Long but blunt claws are visible. Typical two-footed bounds result in sets of two tracks 40–60 cm apart. **Feeding signs:** If feeding on toads, does not eat the head as it is poisonous. Leaves a large, jagged, roughly square opening when feeding on eggs.

Habitat: Occurs in lowland areas with a mixture of open ground and woodland, usually with dense cover; often near rivers and wetlands. In winter, often hunts near farm buildings for rats and mice. Also regularly found in gardens, under sheds and decking.

Food: A wide variety of carrion and live prey, including rodents, Rabbits, birds and small invertebrates such as earthworms. The Polecat is well known for preying upon spawning frogs and toads, often killing more than it can eat, leaving multiple corpses, or paralysing them with a bite to the neck, for storage and later consumption. It can kill animals as large as geese or Brown Hare.

Habits: Active by night and at dusk throughout the year, although less so in mid-winter. May excavate a den, but usually takes over a natural crevice or Rabbit burrow. The den is occupied by a single animal (or a female with kits), which may have several dens and commute regularly between them. Regularly travels 3–4 km in the course of a night's foraging; individuals have a home range of up to 500 ha, but are not particularly territorial, with little aggression or even scent-marking.

Breeding behaviour: Mating is a vigorous process: the male grabs the female around the neck, drags her around to stimulate ovulation, and then copulates for up to an hour. Gives birth in March–June to 3–7 kits (occasionally more). The kits are born with silvery fur, which darkens by the time they are weaned, about four weeks later.

Population and status: British population estimated at more than 45,000, now returning to much of its former range following reduced persecution for fur or to protect poultry and game birds. Recovery aided by unofficial reintroductions, despite some questions about its native status. Escaped Ferrets may occur anywhere, and there are several established feral populations where the Polecat is absent, notably Ireland, the Isle of Man and some Scottish islands.

Stoat

Mustela erminea

(Ermine)

One of the world's most widespread predators, the Stoat occurs across Eurasia and North America and is common throughout Britain and Ireland. It is very hardy, surviving even Arctic winters, when (as in northern parts of Britain) it may moult into a completely white winter coat, called ermine. Exclusively and fiercely predatory, Stoats are capable of tackling prey much larger than themselves.

Fairly common	
HB:	27–40 cm (male), 19–25 cm (female)
T:	7–20 cm
Wt:	100–450 g (male), 50–280 g (female)

Identification: A long, slim, short-legged predator, the Stoat is considerably larger than the otherwise similar Weasel (*page 224*), and can be distinguished by the black tip to its much longer, bushier tail. Brown above and white below, the divide between colours is clean and straight (ragged and irregular in the Weasel), although often less so in Ireland, where Weasel is absent. Other mustelids are larger, or lack the brown/white contrast. Stoats from Scotland and the north of England may moult in winter into a fully white coat, except for the black tail-tip; elsewhere, any seasonal change leads at most to a piebald appearance.

Sounds: Rarely heard, but will make short, sharp shrieks.

Signs: Subtle and hard to find but presence can be confirmed by droppings and tracks. Large prey animals are dragged to a sheltered place to be eaten, and carcasses may show bite marks on the neck. A Rabbit will often squeal loudly when being attacked by a Stoat. **Droppings:** Quite narrow, thin and twisted, about 4–8 cm long × 0·5 cm wide, often with narrower threads at one end. Scats usually contain fur, and have a strong musty smell when fresh; this fades quickly and the colour turns grey. Larger than those of Weasel but smaller than American Mink (*page 214*) or Polecat (*page 218*), scats are usually deposited singly (*e.g.* on a rock). **Tracks:** small, prints of fore feet 2 cm long × 2·2 cm wide; hind feet 4 cm × 2·5 cm. The five digital pads radiate from a lobed interdigital pad and, if visible, the claw marks are small and sharp. Similar to those of Weasel but slightly larger, and with a longer stride of 30–70 cm. Tracks weave around, not in a straight line.

Habitat: Like the Weasel, occurs in a broad range of habitats, but only along the edges, not the interior, of woodland. Often seen in open places such as moorland, scree slopes, the drier parts of marshland and other wetlands, downland, fields with dry stone walls, and rural farmland. Keeps to cover wherever possible, but may be seen scurrying across tracks and roads, by day and night.

Food: Carnivorous, preying particularly on Rabbits and hares, rodents, Moles and birds. Typically, females take mostly voles (including Water Voles), while males tackle Rabbits and hares; prey is killed by a bite to the back of the neck. Forages by moving around its territory on a zigzag route, seeking prey by sight or smell. It can climb well, and will raid bird nest boxes in spring and summer.

WHERE TO LOOK/OBSERVATION TIPS

Rabbit warrens are attractive to Stoats, so patient waiting nearby may bear fruit eventually. If you encounter a Stoat, it is worth making a squeaking sound by sucking the back of your hand: this may be perceived as a Rabbit's distress call, and the naturally curious Stoat will often approach more closely.

▲ The Stoat shows a neat dividing line between the brown back and white front; in the Weasel this dividing line is more uneven.
▼ In the north, moults into an all-white coat in winter, but retains the diagnostic black tip to the tail.

Habits: Active all year, and at any time of day and night, with periods of rest interspersed with foraging and other activities. Runs with the typical two-footed bounding gait of small carnivores, stopping occasionally and standing up on its back legs to peer about. Individuals are territorial, although the size of the territory varies widely and those of males encompass those of one or more females. Scent, urine and scats are used to mark territorial boundaries, males defending their territory against incursions by other males. Each individual may use several dens, most of which are in the burrows of prey (*e.g.* rats or Rabbits) and lined with the fur of the former occupant.

Breeding behaviour: Mating occurs between April and July, but delayed implantation means that births are in April–May the following year. There is one litter a year, the 5–12 kits being born in a natal nest lined with fur, grass and leaves. The young are born blind and with white down; their eyes open at 5–6 weeks, they are weaned by 7–12 weeks and by 10–12 weeks are able to hunt for themselves. Males will enter a family nest before the young are weaned and mate with all the females present, so both the female and her young leave the nest pregnant.

Population and status: Widespread across Britain and Ireland and often common. Population estimated at 462,000 in Britain. In Ireland, where it is fully protected, the population is estimated at 160,000. Widely persecuted for centuries to protect game-rearing interests.

The long, black-tipped tail separates the Stoat from the superficially similar, but smaller, Weasel.

Particularly active, and more often seen, during the summer when feeding young. Stoats, Weasels and their relatives often adopt an alert, upright posture when searching for prey or if they feel threatened.

Both Stoat and Weasel may occasionally be seen 'dancing' in front of a large prey animal, seemingly to distract it from its immediate danger. However, this strange behaviour, which may include manic leaping and rolling around, might be caused by a parasitic infection.

Weasel

Mustela nivalis

(Least Weasel)

The world's smallest member of the order Carnivora, the Weasel is a specialist predator of voles, its size enabling it to go down the narrow burrows of its prey, and to hunt underground. It is very hardy, capable of living the entire winter under snow, and secretive, most often seen scurrying across a path before returning to the safety of cover.

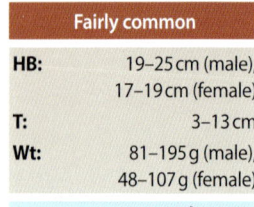

Fairly common	
HB:	19–25 cm (male), 17–19 cm (female)
T:	3–13 cm
Wt:	81–195 g (male), 48–107 g (female)

Identification: Similar to, but markedly smaller than, the Stoat (*page 220*), with short legs, an elongate body, flattened head and large eyes. Also has a distinctly shorter tail than Stoat, which lacks a black tip. It is usually a darker shade of brown and the division between the brown back and white belly is rather uneven, not straight. Never turns white in winter in Britain (although it does in other northern and montane parts of its world range).
Often stands upright on its hind legs, and has a more scampering gait than the bounding gait of a Stoat.

Sounds: Rarely heard in the wild, but hisses, trills and shrieks.

Signs: Subtle and hard to find. The remains of small birds and mammals may be visible outside burrows used by Weasels.
Droppings: Cylindrical, long, narrow, stringy and twisted, with one end thread-like. Usually black, 3–6 cm long, smaller than Stoat droppings, with a musty smell when fresh and usually containing fur. **Tracks:** Extremely small, prints of fore feet 1·3 cm long × 1 cm wide; hind feet 1·5 cm × 1·3 cm. Digital pads with obvious sharp claws radiate from the three-lobed interdigital pad. Stride 25–30 cm; moves with the two-footed bound typical of its family.

The small size and short, plain tail are the key identification features.

Although common, Weasels are difficult to find – usually seen simply dashing across a path. If you do see one hunting, often along hedges or dry stone walls, stand still and wait: they are meticulous hunters, checking likely places for prey thoroughly, and occasionally come out into the open. The sound of mobbing birds may indicate the presence of a Weasel. As with the Stoat, sucking the back of your hand to imitate a Rabbit's distress call may encourage a curious Weasel to approach closely.

Habitat: Occurs in a wide range of habitats: almost everywhere its small mammal prey is found, including farmland and woodland, but scarce in urban areas and not normally seen in gardens.

Food: Preys predominantly on small mammals (60–80% of the diet), particularly voles, but also takes Rabbits and birds, including eggs and nestlings, sometimes raided from nest boxes. A Weasel must eat at least once every 24 hours to survive, and routinely consumes a third of its body weight daily; surplus prey may be killed and cached for later consumption.

Habits: Active year-round and by day and night, alternating rest periods with activity. When hunting, often keeps out of sight in dense vegetation or within the burrow systems of its prey. Very active when foraging, it can travel 1·5 km in an hour. Each individual occupies a home range of up to 25 ha, larger for males than for females, with male territories overlapping female territories but not those of other males: scent-marking avoids most aggressive territoriality. Dens are in the burrows of prey species, including Moles.

Breeding behaviour: Mating takes place mainly from April to July, and young are born between April and August; males are promiscuous and take no part in raising the young (kits). The litter size is 4–6 and the kits are born in a natal nest inside a lined burrow. They are weaned at four weeks of age and can kill for themselves when eight weeks old. A female can rear two broods in a summer if the food supply is good, but may fail to breed at all in poor years.

Population and status: Common and widespread in Britain, but absent from most offshore islands and Ireland; population estimated at 450,000. Not protected and often controlled by trapping, although it rarely attacks gamebirds. Although not specifically protected by domestic legislation, the Weasel is covered by some aspects of EU legislation (see *page 315*).

Generally preys on voles, but will also take mice if the opportunity arises.

Wild Boar

Sus scrofa

Although native to Britain, the Wild Boar was hunted to extinction by about 1300 AD. Those currently living wild in a few parts of Britain originate from escapes and deliberate releases from Wild Boar farms since the 1980s; the main viable populations are in the Forest of Dean, Gloucestershire, and the Sussex Weald. The ancestor of most pig breeds, Wild Boars readily hybridize with domestic pigs and such hybrids may constitute a part of wild populations.

Identification: Usually dark-coloured, hairy pigs with short, stocky legs, a heavy body, big head with a long snout, hairy ears and short tail with a tassel at the end. The shaggy coat forms a prominent ridge down the spine of the male, and becomes thicker in winter. The canine teeth of the male are enlarged into tusks, the longer, lower ones growing to 12 cm and protruding from the mouth. Young piglets are brown with whitish longitudinal stripes. Hybrid pigs are difficult to distinguish, but variation in the colour of the coat, pink on the snout, floppy ears or curly tails are reasonably reliable signs.

Sounds: Pig-like grunts, snuffling and squealing noises; also crashes noisily through woodland.

Reintroduced; uncommon	
HB:	1·0–1·7 m
HS:	70 cm
T:	16–30 cm
Wt:	100–175 kg (male), 80–120 kg (female)

Easily recognised by its shaggy fur, which is thicker in winter.

Signs: The tracks and droppings are distinctive, and it leaves extensive feeding signs and wallows. **Feeding signs:** Roots up pasture and grassland, turning over the earth with its snout and tusks in irregular but sometimes extensive patches. **Droppings:** Sausage-shaped, typically 10 cm long × 7 cm wide, often in clumps. Usually black or dark brown, and have a sweetish smell. **Tracks:** Large, two-toed prints, 5–8 cm long × 4–6 cm wide, with broad, rounded cleaves. Usually shows two small, point-like dewclaw prints, behind and wider apart than the main cleaves. Similar deer

Wild Boar piglets are brown and stripy.

prints do not usually register the dewclaws; if they do, they are farther behind the main track than on a Wild Boar. Stride about 40 cm.
Other signs: Wild Boars wallow in muddy pools, creating large shallow depressions, usually with tracks and scats nearby. They rub the mud off their fur against rocks or tree trunks; regular use of a tree trunk can wear away the bark at a height of 60–80 cm.

Habitat: Found in large woodland complexes with plenty of cover. Often forages in nearby fields, marshes and other open habitats.

Food: Has a broad diet but primarily herbivorous, including nuts and seeds (acorns, chestnuts, Beech mast, *etc.*), fruits, roots, and crops such as maize and potatoes, all taken from or rooted out of the ground. Also takes invertebrates (*e.g.* worms), carrion, the eggs and chicks of birds, and occasionally small mammals.

Habits: Active throughout the year; mainly nocturnal and crepuscular. Roams widely at night over an area of up to 2,000 ha according to food supply, often returning to the same resting site during the day. Several mature females and their recent offspring may gather into informal groups ('sounders') of 6–30 animals; mature males are solitary except during the rut.

Breeding behaviour: The rut (mating season) begins in winter, especially December, although younger sows may mate in March and April. Dominant males are polygamous and prevent younger males from mating. The sow builds a nest 30–70 cm high from nearby vegetation, where she gives birth to usually 4–6 piglets (range 2–10). The mother and her litter re-join the female group after about a week, and youngsters are weaned at 3–4 months. Mothers with young can be very aggressive when threatened.

Population and status: British population 100–500; very small numbers in Ireland. The considerable potential for increase in Britain divides opinion: although a former native species which adds natural dynamism to its habitats, benefiting biodiversity, it can cause considerable damage to crops, gardens and playing fields, for example, and potentially spread diseases to domestic animals.

See also: Tracks *p. 52* | Droppings *p. 43*

Identifying deer

This is a comparative gallery of Britain's two native and four naturalized species of deer. The coat colour of deer is variable, so for identification it is best to concentrate on the rear end (rump) pattern. Antlers are found only on male deer, and are shed every year. Other signs, such as droppings and tracks, are more subtle and experience is required to identify the species concerned.

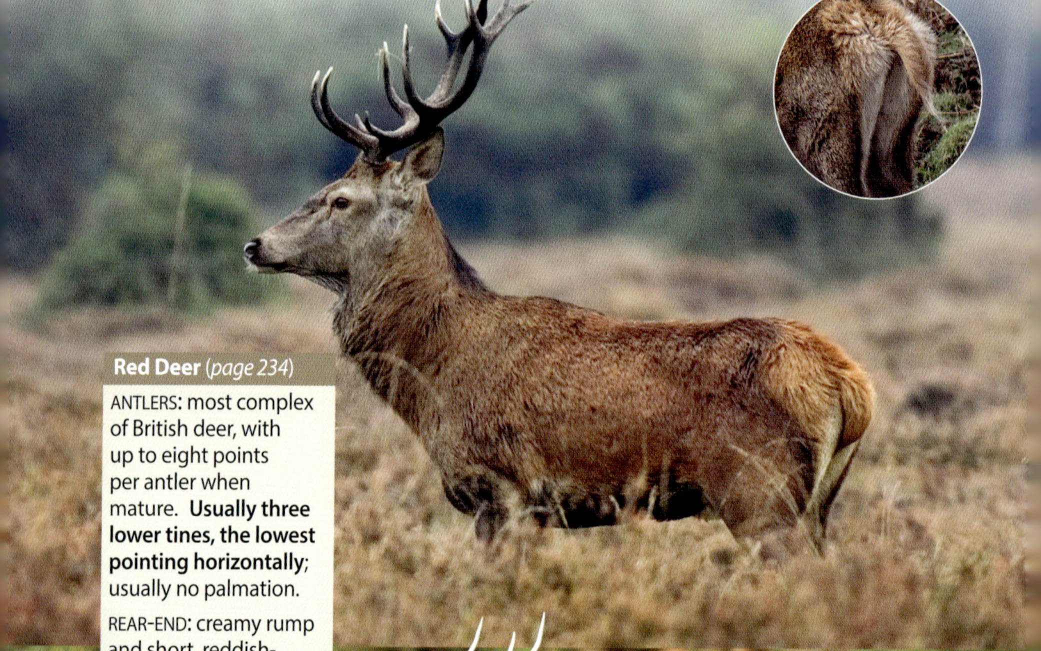

Red Deer (*page 234*)

ANTLERS: most complex of British deer, with up to eight points per antler when mature. **Usually three lower tines, the lowest pointing horizontally**; usually no palmation.

REAR-END: creamy rump and short, reddish-brown, unmarked tail.

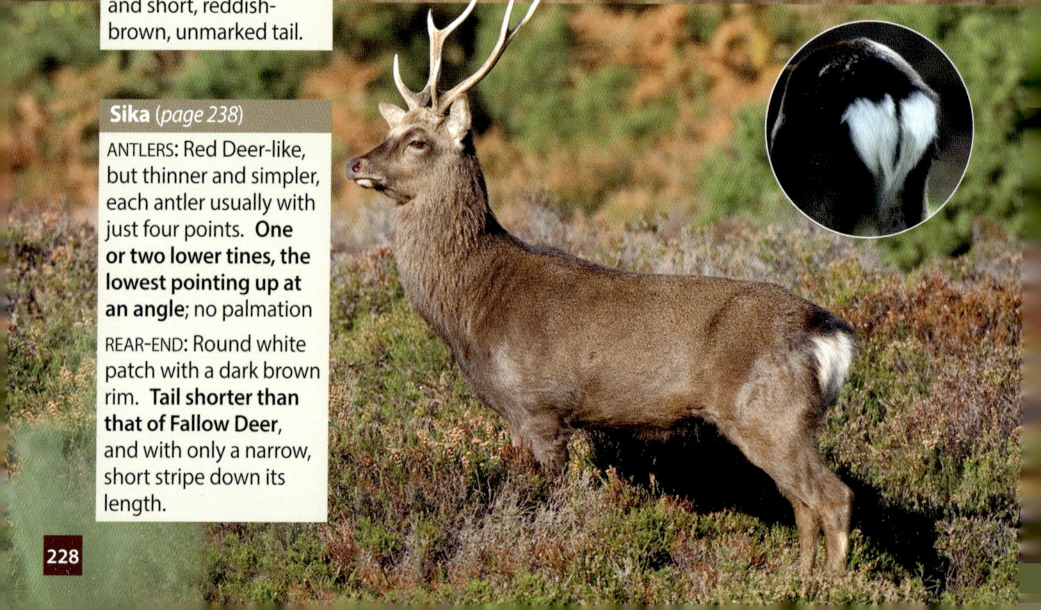

Sika (*page 238*)

ANTLERS: Red Deer-like, but thinner and simpler, each antler usually with just four points. **One or two lower tines, the lowest pointing up at an angle**; no palmation

REAR-END: Round white patch with a dark brown rim. **Tail shorter than that of Fallow Deer**, and with only a narrow, short stripe down its length.

Chinese Water Deer
(*page 248*)

ANTLERS: none.
TUSKS: small tusks (♂), longer than those of Reeves' Muntjac.

REAR-END: no obvious rump patch; short tail grey-brown, the same colour as the back.

Reeves' Muntjac
(*page 246*)

ANTLERS: very short, a single spike, often curved inwards.
TUSKS: small tusks (♂).

REAR-END: no obvious rump patch: backside and short tail reddish brown. When alarmed, raises the tail to show a brilliant white underside.

European Roe Deer
(*page 230*)

ANTLERS: short, upright, not much taller than head. Crusty at base, and just two or three points.

REAR-END: broad, whitish rump patch, kidney-shaped (♂) or inverted heart-shaped (♀). No obvious tail.

Fallow Deer
(*page 242*)

ANTLERS: **unmistakable, broad-tipped, palmate**.

REAR-END: oval white patch with a black rim. Bisected by **long tail**, which usually has a broad black stripe down the middle; white below.

229

European Roe Deer   *Capreolus capreolus*

(Roe Deer, European Roe)

The European Roe Deer is one of just two native deer in Britain, along with the much larger Red Deer (*page 234*). Although now our commonest and most widespread deer, historical persecution and deforestation meant that by the 18th century it had been extirpated from most of the country, apart from a remnant population in the Scottish Highlands. It is thought that most European Roe Deer in England now are derived from deliberate reintroductions.

SOME PROTECTION (UK)	
Common	
HB:	0·95–1·25 m
HS:	60–75 cm
T:	2–3 cm
Wt:	10–25 kg
Male slightly larger and heavier than female	

Identification: An elegant small to medium-sized deer, about the size of a goat. Appears almost tail-less, with a large whitish or buff rump patch, shaped like an inverted heart; the female (doe) in winter has a tuft of white hairs that looks superficially like a tail. Adults have a plain coat without any spots at any time (unlike the similar-sized Fallow Deer (*page 242*) or Sika (*page 238*)), reddish-brown in summer but darker grey-brown in winter. The short muzzle has a black nose and moustachial streaks, which contrast with the white chin. Has a longer neck and more pointed ears (outlined in black) than the similar but smaller Reeves' Muntjac (*page 246*), and lacks that species' typical hunched back. Fawns have white spots for the first few months.

Antlers: Males only. In contrast to other deer (which shed antlers in spring), the antlers are shed between October and December; they immediately start to re-grow, completing by March. They are small and upright, up to 30 cm long, ridged at the base and set close together on the head Each has just three points: one basal and two at the top.

Sounds: Makes a loud, dog-like bark, especially at twilight and also when flushed.

Signs: Include tracks and droppings, and damage to vegetation. Eyeshine is white/light green, about 1 m above the ground. **Tracks:** Small, two-toed prints, 4·5 cm long × 3·5 cm wide. The cleaves are slender and parallel, forming a narrow heart shape sharply pointed towards the tip; both the outer walls and toe pads are prominent. When the animal is running, the cleaves may be splayed. Stride 40 cm when walking; 60 cm when trotting. **Droppings:** Cylindrical, 1·4 cm long × 0·8 cm wide, finely pointed at one end with a dimple at the other. In winter they are black, shiny and separate, while in summer they are often browner, more fibrous and adhere into clumps. Droppings may be scattered or clustered in latrines. **Signs on trees:** Browsing of shrubs and saplings may create a browse line at about 1·2 m from the ground. In summer bucks rub their forehead against small trees, creating marks 10–80 cm above ground, and breaking off smaller twigs and branches. **Rutting scrapes:** Males in rut scrape the ground with their hooves, creating a bare patch. During courtship a doe may run repeatedly around a small tree, making a ring of trampled vegetation. **Bedding sites:** Unlike other species, clears the ground for bedding sites, scraping away leaves or branches.

WHERE TO LOOK/OBSERVATION TIPS

Shy, but common and not difficult to see in favoured areas, especially at dusk and dawn. Often grazes on the edge of fields, close to woodland, even during the day.

▲ Male (buck). The black nose and white lips and chin are characteristic of both sexes.
▼ Female (doe). Does are 5–10% lighter in weight than bucks.

Fawns are often concealed among leaf-litter ...

... but are able to run when just a few days old.

Habitat: Mixed landscape of woodland and fields, favouring deciduous woodland with wide rides and clearings. Sometimes also in more open areas, including moorland, particularly in Scotland.

Food: Browses trees, shrubs and herbs, including brambles, Ivy, Rosebay Willowherb, and Heather, to a height of 1·2 m. Also feeds from the ground, taking fungi, acorns and other fruits in autumn and winter, and eats cereal crops and garden plants, sometimes causing damage.

Habits: Active all year and throughout the day, but particularly so at dawn and dusk; at other times alternates short periods of rest with feeding. It is often solitary, sometimes in family units and, especially in winter, larger groups. Between March and September males hold small, defended territories (typically 5 ha), in which they scent-mark copiously on the ground and on small trees. If challenged they will bark and chase, and sometimes lock antlers with one another. These territories break down over the winter. The females inhabit home ranges that are not defended and often overlap. Youngsters disperse over a distance of only a few kilometres. When startled, makes a bounding escape through the undergrowth or over fields.

Breeding behaviour: The rut is from mid-July to the end of August, earlier than most other deer. Bucks detect fertile females by scent, mostly within their own territory, although towards the end of the rut they may search more widely. A long chase ensues, lasting on and off for 2–5 days, in which the male follows the female in a circle or 'figure-of-eight' that often becomes trodden down (a 'roe ring'). After mating, implantation is delayed until late December/early January, and two (occasionally three) young are born in May/June. The fawns are suckled for 6–10 weeks or more, are typically kept about 20 m apart for safety, and remain with their mother until the following summer.

Population and status: British population about 500,000, probably increasing. Absent from Ireland. Under The Deer Act of 1991, this species cannot be shot during the 'Close Season' from 1st November to 31st March inclusive (buck) or 1st April to 31st October inclusive (doe), or any time at night from an hour after sunset until an hour before sunrise.

When running away, does show a 'tufted' white rump.

▲ Well-grown youngsters of both sexes resemble does.
▼ A buck European Roe Deer 'in velvet' with a doe; the grey spot on the throat is characteristic.

Red Deer   *Cervus elaphus*

Britain's largest native land mammal, the Red Deer is renowned for its impressive multi-pointed antlers and the spectacle of its autumn rut. Although mainly a forest deer in mainland Europe and in England, it is most closely associated in Britain with the hills and moorland of the Scottish Highlands. First introduced into Ireland about 5,000 years ago, most of the current small population there stems from releases in the 17th and 18th centuries.

SOME PROTECTION (UK)	
LEGALLY PROTECTED (Republic of Ireland)	
Locally common	
HB:	1·6–2·6 m (male), 1·7–2·1 m (female)
HS:	1·14–1·22 m
T:	12–15 cm
Wt:	130–300 kg (male), 80–150 kg (female)

Identification: A tall, straight-backed, reddish-brown (darker in winter) deer with a distinctive creamy-yellow rump. The head is long, with a tapered muzzle, and the ears are long and relatively narrow. Stags are heavily built, with antlers (in season) and a shaggy neck, whereas hinds are smaller and slimmer. Sika (*page 238*) and Fallow Deer (*page 242*) have white rump patches at least partly outlined with black; European Roe Deer (*page 230*) is a similar colour to Red Deer, but is much smaller and has a shorter, blunter muzzle. When very young, Red Deer calves are usually russet with white spots, and have a longer head than the similarly spotted European Roe Deer, Sika and Fallow Deer. However, the spots are lost after about two months.

Antlers: Males only; shed March–April, fully grown again by August. Larger than those of other British deer, up to 1 m across, and in older males may have 8 points to each antler. Usually has three lower points (tines) branching from the main beam, the lowest of which projects horizontally before bending upwards; Sika usually has two lower tines, the lowest projecting upwards, while the antlers of Fallow Deer are broad and palmate. A young male Red Deer has smaller antlers, often in the form of a single, curved spike. Can occasionally have partially palmate antlers, although such individuals are more prevalent in herds that are confined to parklands than in wild populations.

Sounds: Stags in rut make a loud, drawn-out, bellowing roar, audible over a distance of several kilometres.

Signs: Include tracks and droppings, with a variety of other signs including vegetation damage, ground disturbance and wallows. **Tracks:** Two-toed prints: hinds 6–7 cm long × 4·5–5 cm wide; stags 8–9 cm long × 6–7 cm wide. The cleaves are rather parallel without much splaying, relatively broad, and with a convex outer edge. The edges and toe pads are well defined, but the dewclaw is rarely visible. Stride when walking approximately 1 m, and when trotting 3 m. **Droppings:** Cylindrical, 2–2·5 cm long × 1·3–1·8 cm wide, finely pointed at one end and with

WHERE TO LOOK/OBSERVATION TIPS

Easy to find in the Scottish Highlands and on some islands, such as Rum, Islay and Mull. To see and hear the rut in autumn is one of the most exhilarating British mammal-watching experiences. In England, try Exmoor, the Quantocks, the New Forest, Cannock Chase and the Peak District, but all populations are rather local within these areas. Generally trickier to find in forest habitats, but deer parks (*e.g.* Richmond Park in London) offer a good view of the rut.

▲ Males (stags) are heavily built and often have a shaggy neck.
▼ Female (hind) with young (calf). Note the even reddish colour and the straight back.

a depression at the other. Black and glistening at first, then dull brown. Often stuck together in clusters or strings. **Signs on trees:** In woodland, creates a browse line at 2 m or more (see *page 45*). Strips bark up to about 2 m in winter and spring, sometimes leaving tooth marks; also rubs its body against trees and marks them with antlers, especially during the rut. **Wallows:** Deep, waterlogged depressions in the ground, often an enlarged existing puddle; dark brown hairs are usually scattered in and around the wallow. **Rutting signs:** Scrapes the ground with its hooves, sometimes forming a hollow in which it rolls to leave the scent of bodily fluids. Often rubs itself against trees near the rutting stand, and leaves scratches on the trunk, up to a height of 2 m.

Habitat: There are two key habitats in Britain: moorland, where Red Deer can live year-round in treeless terrain; and woodland, including both coniferous and deciduous. In open country, stags are often found in less productive heather moorland, while hinds are usually in more fertile patches, with grassland.

Food: Grass forms the most important dietary component throughout the year. Also browses Heather and other ericaceous shrubs (males especially), brambles, Holly, Ivy, Bracken, bark and the shoots of trees and shrubs.

Habits: Active throughout the year and at any time of day and night, although particularly around dusk and dawn. In the uplands, Red Deer move between the upper slopes for daytime foraging and lower elevations at night, while forest-dwelling animals visit nearby open grazing areas, preferentially in darkness. Outside the rutting period, males and females do not mix: hinds gather into groups of related females, while stags form transient groups of unrelated animals. Such groups can be large, particularly on mountains and moorland, and have an internal hierarchy, the oldest and largest animals being dominant. Individuals have home ranges of 200–2,400 ha, with one or two core areas of activity. Young females do not wander far from their natal area, but stags may travel widely for their first few years of life.

Breeding behaviour: Mating occurs during the rut, from late September to early November. Individual stags leave their male herds and go to a special rutting area, which is often traditional over the years. They round up hinds into a harem (up to 15 females, fewer in woodland habitats), which may then become the focus of competition with a few other males (other non-competing males simply remain in bachelor groups nearby, usually within sight of the harem). Harem-holding males roar loudly (largest males roar the most), thrash vegetation with their antlers, and spray body fluids liberally, without any provocation. If challenged, the roaring contest descends into a showdown, when stags first walk alongside one another and then lock antlers in their famous contests, the striking of which can be heard at a great distance. Fights often result in injury and sometimes death. Once mated, delayed implantation means that a hind will not give birth until the following spring, usually from mid-May to the end of July (sometimes later). The single calf is born in a quiet hiding place away from other animals, but soon leaves to accompany the mother and may then mix with other youngsters. It is fully grown and able to breed at about 15 months; stags do not usually breed until 5–6 years old.

Population and status: Around 500,000 in Britain, stable or increasing (especially in the lowlands); about 800 in Ireland. Some populations are culled or hunted for sport or venison. Interbreeds freely with Sika, which threatens the genetic integrity of some native Red Deer populations. Under The Deer Act of 1991, this species cannot be shot during the 'Close Season' from 1st May to 31st July inclusive (stag) or 1st April to 31st October inclusive (hind), or any time at night from an hour after sunset until an hour before sunrise.

▲ Although a Red Deer stag's antlers become larger and increasingly impressive as the animal matures, it is rarely possible to age an individual accurately by this means.

▼ Stag, hind and calf: note the creamy-yellow rump, in contrast to the white of Sika and Fallow Deer.

Sika

Cervus nippon

(Sika Deer)

SOME PROTECTION (UK)
LEGALLY PROTECTED
(Republic of Ireland)

Introduced; locally common

Introduced to Britain from Japan in the 19th century, the Sika has become naturalized in the wild as a result of escapes from captivity and deliberate releases. Well-established populations are now found in a number of regions, including south Dorset, the New Forest, Lancashire, the Lake District, Killarney, and parts of the Borders and Highlands of Scotland. It is very closely related to the Red Deer and the majority of British populations are probably hybrids.

HB:	1·2–1·9 m (male), 1·1–1·6 m (female)
HS:	1·07–1·22 m
T:	10–15 cm
Wt:	40–55 kg (male), 28–40 kg (female)

Identification: Similar to Red Deer (*page 234*), but smaller and more slender, with a distinct white gland on the lower back leg, and always distinguished by the white rump neatly delineated above by a black border: Red Deer has a cream patch without the black border. The tail is shorter than that of a Fallow Deer (*page 242*), and mostly white, with just a small streak of black that does not extend right down the tail. In summer, it can be distinguished from Red Deer by its spotted coat. In winter, after the rut, the coat may show few if any spots, but is darker than that of Red Deer. European Roe Deer (*page 230*) is smaller and lacks spots, has no black on the rump and a much shorter tail.

Antlers: Males only; shed in April–May, the velvety covering of newly-growing antlers remains until August–September. Although impressive, they are smaller than those of Red Deer of a comparable age, with at most four points per antler; they always lack the second branch off from the main beam shown by Red Deer. Fallow Deer have palmate antlers and those of the Roe Deer are much smaller.

Sounds: Stags in particular make a variety of grunts and groans, and during the rut produce high-pitched whistles and screams.

Signs: Tracks, droppings and tree damage, often hard to distinguish from those of other deer. **Tracks:** Two-toed prints, up to 6 cm long, have sharp outlines as toe-pads are barely visible. More sharply pointed than that of a Red Deer, and the cleaves are not as broad or as curved. Difficult to distinguish from Fallow Deer tracks, but broader and more splayed; smaller and less splayed (especially the front print) than those of European Roe Deer. **Droppings:** Smaller than those of Red Deer, to 1·5 cm long, finely pointed at one end and flat or indented at the other. Black and glistening at first, then dull brown. Deposits droppings in latrines more often than do other deer. **Signs on trees:** Scratches antlers on trees during the rut, leaving deep vertical grooves, more so than other deer. **Rutting signs:** Scrapes a pit in the ground, like that of Red Deer.

Habitat: Coniferous and deciduous woodland next to heathland, typically on acid soils. Unlike Red Deer, generally avoids treeless locations, although sometimes feeds in bogs and saltmarshes.

WHERE TO LOOK/OBSERVATION TIPS

Its more solitary and crepuscular habits make Sika generally harder to find than Red Deer. However, it is easy to see at Arne in Dorset, where it feeds in the open on the salt-marshes.

▲ Male (stag) in spring coat
▼ Female (hind) in spring coat

Food: Browses and grazes; mainly grasses and Heather, along with shoots and leaves of coniferous and deciduous trees, young Holly, gorse, bark and acorns.

Habits: Active year-round and throughout the day, with peaks at dawn and dusk; if disturbed, tends to lay low. Not particularly sociable, many animals are largely solitary when not breeding. However, small groups of hinds and young sometimes form, and animals may also come together in favoured breeding areas. Such herds can be especially damaging to sensitive habitats due to the impact of trampling. As with Red Deer, the sexes remain separate until the rut. Individuals live in a small home range, typically 18–22 ha for hinds and 45–55 ha for stags (young stags larger).

Breeding behaviour: The rut lasts from the beginning of October to late November, although stags may sometimes give their characteristic call well outside these times. Their mating strategy varies: some stags adopt a harem, others hold a rutting territory and still others wander in search of receptive females and mate opportunistically. A single calf is born between early May and late June; it is fed on milk for six months, but begins to take solid food when about 10 days old.

Population and status: Estimated to be 11,500 in Britain and up to 25,000 in Ireland; increasing, although most have a genetic influence from Red Deer. It is believed that the only pure-bred Sika may be in the New Forest, the Scottish Borders and around Killarney, Ireland. Under The Deer Act of 1991, this species cannot be shot during the 'Close Season' from 1st May to 31st July inclusive (stag) or 1st April to 31st October inclusive (hind), or any time at night from an hour after sunset until an hour before sunrise.

Hinds in winter coat; the black-edged white rump is characteristic.

▲ A Sika stag in summer coat (the dark markings are the result of wallowing in mud).
▼ Hinds in summer coat, and calf.

Fallow Deer   *Dama dama*

Brought to Britain from the eastern Mediterranean region in the 11th century for hunting, the Fallow Deer is now the most widespread and abundant of our introduced deer, found throughout England and more patchily in Wales, Scotland and also in Ireland, where it was introduced in the 13th century. It is the deer most commonly kept in parks and, being sociable, often forms substantial herds.

Identification: A medium-sized deer; mature males (bucks) have unmistakable flattened, palmate antlers between autumn and spring. The rounded white rump patch is bordered with a black 'horseshoe', and bisected by the rather long tail, which is usually black above and white below. In the similar-sized Sika (*page 238*), the tail is mainly white, with only a short narrow black stripe, and the black border encircles the white rump patch. European Roe Deer (*page 230*) has a creamy-white rump and no obvious tail. The buck Fallow Deer has a noticeably hairy penis sheath. In summer, the coat is a rich reddish-brown with numerous white spots and a white horizontal line on the flank (this latter feature lacking in Sika); it is duller brown and spotless in winter. However, the coat colour is variable: herds often contain entirely dark or creamy-white individuals, which show little colour variation between the seasons.

SOME PROTECTION (UK)	
LEGALLY PROTECTED	
(Republic of Ireland)	
Introduced; common	
HB:	1·45–1·55 m (male), 1·30–1·45 m (female)
HS:	70–95 cm
T:	10–15 cm
Wt:	50–80 kg (male), 35–50 kg (female)

Antlers: Males only; shed in April, and fully grown again by August. Uniquely among British and Irish deer, the antlers are palmate, flattened towards the tip with several sharp points. Yearlings have spiked antlers, similar to those of Sika.

Sounds: During the rut, bucks produce a series of loud belching and groaning noises, similar to those of Red Deer (*page 234*) but less drawn out.

Signs: Droppings and tracks are difficult to distinguish from those of other deer, and sometimes also sheep. Bucks scratch their antlers on trees in summer to remove the velvet, often leaving remnants and marks on the bark. **Droppings:** Typically black and cylindrical, with one end pointed and the other indented, about 2 cm long × 1·5 cm wide; usually deposited in piles or clumps, with some pellets fused. **Tracks:** Two-toed prints, 6·5 cm long × 4 cm wide. The cleaves are close together and not usually splayed, narrow, with a sharp point and almost straight outer edges at the rear. In soft mud the dewclaws are visible, set well back. Smaller and more elongated than Red Deer tracks, narrower than those of Sika and larger than European Roe Deer tracks. The stride length when walking is about 60 cm. **Rutting signs:** Bucks scrape the ground with their hooves, especially beneath trees, and may tread in droppings. **Feeding signs:** Strips bark off trees, especially in spring, to a height of 1·4 m or more.

Habitat: Deciduous woodlands, copses and adjacent fields. Often kept in open parkland, with grassland and mature trees.

WHERE TO LOOK/OBSERVATION TIPS

Common but localized; generally easy to see where it occurs, especially in herds within deer parks.

▲ Male (buck) in spring; the palmate antlers are unmistakable.
▼ Female (doe) in summer, when the coat is usually covered with white spots.

Food: Although principally a grazer, it also takes acorns and other nuts from the ground, and will browse some broad-leaved trees and shrubs such as Holly and Heather, especially when grazing is poor.

Habits: Active throughout the year. Usually largely diurnal, feeding in sunlight and resting in shade, but in disturbed areas tends to be more crepuscular and nocturnal. Extremely sociable, often found in large herds that can number more than 100. Females (does) and males (bucks) often keep apart in bachelor groups and family groups of does and young, except during the rut. Home ranges are small (50–100 ha) and overlap extensively, with little territoriality.

Young (fawns) are born in early summer.

Breeding behaviour: The rut takes place in October and early November. The bucks' mating strategies are diverse, although most establish their own separate rutting stands and attempt to attract females by calling. However, in some cases several males hold territories that overlap, and in more extreme situations groups of 5–25 bucks will gather together on open ground and attempt to attract females collectively in a 'lek', each holding a compact territory of just a few square metres. Some bucks gather a harem of does, while others wander in search of receptive females and mate opportunistically. The female gives birth to a single fawn in May or early June, which walks with its mother after ten days and is weaned at 8–9 months.

Population and status: Current British population is about 100,000; the number in Ireland is unknown. Stable or increasing, many populations are culled to reduce damage to agriculture and forestry, and for venison. Under The Deer Act of 1991, this species cannot be shot during the 'Close Season' from 1st May to 31st July inclusive (buck) or 1st April to 31st October inclusive (doe), or any time at night from an hour after sunset until an hour before sunrise.

A buck moulting into its duller winter coat.

Fallow Deer does: both sexes can vary in colour from white to dark grey, and the coat may not be spotted. Although the tail is always long, not all individuals show a black line down the upperside.

Reeves' Muntjac   *Muntiacus reevesi*

(Muntjac)

A native of China, Reeves' Muntjac was first introduced to Woburn Park, Bedfordshire, in 1901. Subsequent escapes and releases led rapidly to a free-living population. It is now well established over much of central and eastern England, and spreading rapidly to the west and south-west, and northwards. It was first recorded in Northern Ireland In 2000, and has since been confirmed to occur elsewhere in Ireland.

Introduced; locally common	
HB:	90–100 cm
HS:	45–52 cm
T:	15 cm
Wt:	12–15 kg (male heavier than female)

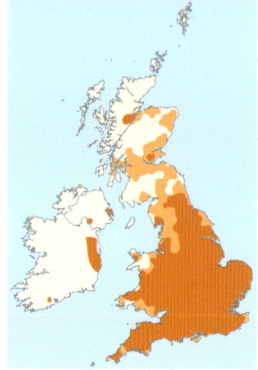

Identification: A small, stocky deer with a characteristic hunched appearance, having an arched back and often holding its head down low, creating a pig-like impression. It has a diagnostic face pattern, with two black stripes running down the forehead that meet in a 'V' between the eyes, and dark pits in front of the eyes. The fur is reddish-brown in summer, but a duller grey-brown between September and April. The tail is brown above but brilliant white below, a feature that is displayed when fleeing from danger, as the tail is held erect. Males have protruding canines, although rarely as visible as those of Chinese Water Deer (*page 248*). Fawns have white-spotted coats.

Antlers: Males only: short (to 10 cm) simple spikes (sometimes with a hooked point), arising from enlarged bony pedicles. They are shed in April–May and grow back through the summer, losing velvet by August–October.

Sounds: A loud, staccato bark, produced at any time, but repeated regularly during courtship.

Signs: As well as droppings, tracks, and rutting scrapes, makes well-worn trails through vegetation 15–20 cm wide. **Tracks:** Small, about 3 cm long × 2 cm wide, one cleave always longer than the other. The stride is 25–30 cm. **Droppings:** Black and shiny, about 1 cm × 1 cm, spherical or slightly cylindrical, rounded at one or both ends. They are usually deposited in heaps. **Rutting scrapes:** Scrapes the ground with its feet to clear an area about 30 cm wide. **Feeding signs:** Creates a browse line at 60–80 cm height, a little lower than that of European Roe Deer (*page 230*). Browses saplings, and may bite through shoots incompletely to reach higher foliage.

Habitat: Secretive, living in thick undergrowth, usually within woodland (both deciduous and coniferous). Also found in larger gardens.

Food: A browsing species, taking a variety of leaves and flowers, including brambles, Raspberry, Ivy, Bluebell and Primrose. This selective browsing can adversely affect the diversity of woodland ground flora, and threaten the survival of some key species, such as Oxlip. The diet varies with the seasons: nuts and fungi are important in the autumn and winter, and it sometimes grazes grasses, especially in early spring.

WHERE TO LOOK/OBSERVATION TIPS

Most common in the Home Counties and East Anglia, where it comes into gardens and urban parks, even during the day. Also sometimes seen feeding along road verges.

Habits: Active year-round, and at any time of day or night, with peak feeding times at dawn, dusk and midday. It is a secretive, shy and solitary animal with a home range of 20–28 ha (males) or 11–14 ha (females). Both sexes are territorial: males (bucks) fight, using their antlers to wound or unbalance an opponent, while females (does) chase one another. Territories are also marked by scent, using faeces or urine, and especially secretions from glands on the face which are rubbed against a tree or the ground.

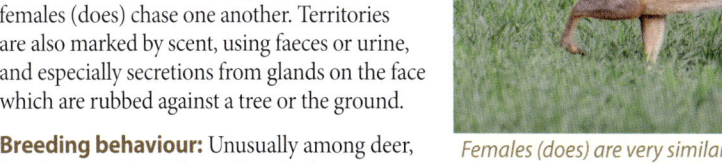

Females (does) are very similar but lack antlers.

Breeding behaviour: Unusually among deer, breeds at any time of year. Females become fertile soon after giving birth and come into oestrus every 14–15 days. Bucks mate with any does that overlap their territory, sometimes several individuals over the course of a year, following a prolonged courtship chase, during which the doe often calls continuously. The single fawn, born in any month of the year, is weaned at four months and may sometimes breed in its first year.

Population and status: British population is at least 52,000, mostly in England; still increasing, especially in the south-west. Where common it causes problems in coppice woodland (affecting the structure and ground flora) and in gardens. The population in Ireland is currently unknown. Although not specifically protected by domestic legislation, the Reeves' Muntjac is covered by some aspects of EU legislation (see *page 316*).

Male (buck)

Chinese Water Deer *Hydropotes inermis*

Native to eastern China and the Korean peninsula, this deer was introduced to Woburn Park, Bedfordshire at the end of the 19th century. Escapes and releases from that and other collections resulted in the establishment of a wild population in England.

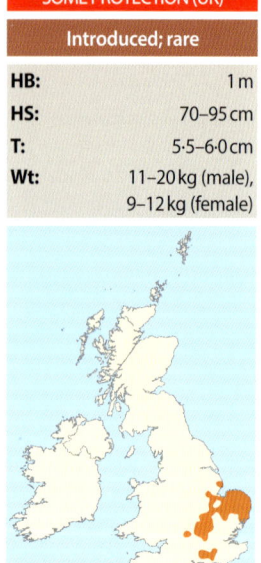

SOME PROTECTION (UK)	
Introduced; rare	
HB:	1 m
HS:	70–95 cm
T:	5·5–6·0 cm
Wt:	11–20 kg (male),
	9–12 kg (female)

Identification: A small deer with no antlers and a creamy-brown winter coat, tinged chestnut in summer; lacks the pale rump patch of other deer species. Adult male has tusks (protruding upper canines) up to 8 cm long. Both sexes have dark, beady eyes and large rounded ears, which are usually held erect, giving a rather 'teddy bear'-like look. Smaller than Eurasian Roe Deer (*page 230*) but slightly larger than Reeves' Muntjac (*page 246*), with a straighter, less hunched back than the latter species. When running fast, it may fling up its hind legs, like a hare. Fawns have dark reddish-brown coats with white spots.

Antlers: None.

Sounds: A loud, growling bark when alarmed, longer than the equivalent Reeves' Muntjac call. Males in rut also utter squeaking sounds.

Signs: Tracks and droppings, but also makes obvious flattened trails through reedbeds. **Tracks:** The cleaves are narrow and pointed, about 4–5 cm long, with a wide gap between them. The inner edge is usually slightly convex. Stride about 35 cm when walking. **Droppings:** Small, 1–1·5 cm × 0·5–1 cm, cylindrical, non-sticky, pointed at one end and rounded at the other. More elongated than those of Reeves' Muntjac, and usually black. Often found in reedbeds.

Habitat: Mainly freshwater marshland, with a combination of reedbeds and woodland; less commonly seen in farmland.

Food: Grazes grasses, sedges and herbs; also feeds on carrots, potatoes and winter wheat.

Habits: Active throughout the year and any time of day or night, with peaks at dawn and dusk. Solitary, seldom forming any kind of group except for mothers with recent young, or when food is scarce. Males are territorial all year and can be extremely aggressive in defence of their territory; these territories are often small (maximum 20 ha). Males in combat first walk alongside one another in threat, then attack head-on with their tusks, often inflicting severe damage. They scent-mark with faeces, urine or secretions from glands.

Breeding behaviour: Differs in many ways from other deer species. The rut is in December. Gives birth to up to five fawns (usually three or fewer), in May–July. The fawns mature very quickly; they are weaned after just three weeks (lactation may last 4–5 months) and are sexually

WHERE TO LOOK/OBSERVATION TIPS

Generally secretive and difficult to see; best searched for by viewing from wetland bird hides in its strongholds, especially Woodwalton Fen in Cambridgeshire, and the Norfolk Broads. However, in the Woburn area of Bedfordshire it can easily be seen by the roadside.

mature at just 6–7 months. Polygamous males mate with any females in their territory, the females being fertile for just 24 hours when in oestrus.

Population and status: British population about 1,500 (plus 600 in parks), slowly increasing in number and range. Strongholds are East Anglia, the Cambridgeshire fens and Bedfordshire, with scattered records in the Midlands and southern England. In its native range it is threatened and categorized as Vulnerable: if current trends continue, it is likely that Britain will soon hold more than half of the world population, posing an interesting conservation conundrum. Under The Deer Act of 1991, this species cannot be shot during the 'Close Season' from 1st April and 31st October inclusive, or any time at night from an hour after sunset until an hour before sunrise.

Females (does) are very similar but lack tusks.

Male (buck)

▲ Summer coat
▼ Coat much paler in winter than in summer

VU Reindeer

Rangifer tarandus

The Reindeer was native in Britain during the Late Glacial (until about 8,000 years ago), but became extinct as the climate warmed up. Since 1952, two small, free-ranging but heavily managed herds have been maintained in the Cairngorms.

Identification: A large, heavily built deer with large antlers. The coat is generally a soft, creamy grey-brown, often with much darker chocolate-brown legs and rear, especially in summer. The male has long, shaggy hairs hanging down from the neck.

Antlers: Present in both sexes. They are long and curved, and branched at the end. The basal tines grow forward and branch, while the main beam grows backwards and upwards before branching from a flattened base. These are shed in December–January (males) or May (females).

Sounds: Makes grunting sounds. When walking, a tendon in the foot makes a clicking sound.

Habitat: Moorland above the tree line.

Food: Feeds naturally on lichens (especially in winter), grasses, sedges, Heather and dwarf shrubs; herds are also provided with supplementary food.

Semi-domesticated	
HB:	1·05–1·60 m
HS:	1·07–1·22 m
T:	10–15 cm
Wt:	40–55 kg (male), 28–40 kg (female)

Habits: Active throughout the year and at any time of day. Forms long-lasting herds, which in native populations are sometimes migratory. Can run at 60 km/h.

Breeding behaviour: The rut lasts from mid-September to early October. A single calf is born in May; it can walk within an hour and grows quickly, fed on exceptionally rich milk.

Population and status: There are about 150 animals in two herds, which are managed but free-ranging on the mountains. Although not specifically protected by domestic legislation, the Reindeer is covered by some aspects of EU legislation (see *page 316*).

WHERE TO LOOK/OBSERVATION TIPS

The Reindeer Company runs commercial trips from the bottom of Cairngorm to get close to one of the herds (cairngormreindeer.co.uk), although the deer can often be seen from the road up to the funicular railway, or from the mountain itself. The other herd is 50 km away on the Glenlivet Estate.

Feral Goat   *Capra hircus*

Introduced from the Mediterranean region in Neolithic times, the earliest recorded remains of goats are some 4,500 years old. Probably from the outset, some individuals escaped or were released and formed feral herds; free-living Feral Goats remain a feature of many rocky landscapes, largely in the uplands and on islands. Present-day Feral Goats often show characteristics of old domestic breeds.

Introduced (prehistoric); locally common	
HB:	1·05–1·60 m
HS:	1.07–1.22 cm
T:	11–13 cm
Wt:	45–90 kg (male), 25–55 kg (female)

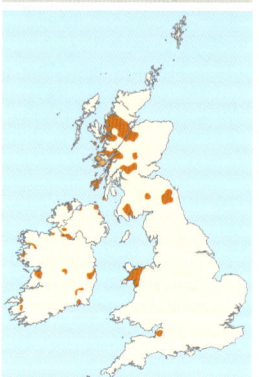

Identification: Stockier and shorter-legged than deer, with shaggy coats (thickest in winter) and backward-pointing horns. Males have larger and longer horns than females, and a beard-like tuft on the lower jaw. There is much variation in the coat colour and pattern, even within a single flock. Feral Goats have erect ears, unlike many domestic forms. Most similar to some old breeds of sheep, but differ in having a beard (males); a flat, long tail; and a callus on the knee.

Horns: Both sexes have horns, which are not shed (unlike deers' antlers); females have shorter, thinner horns than males. The horns grow up and back before diverging outwards; they have raised rings along their length, and often a prominent keel down the front.

Sounds: Makes a bleating contact call, and also a snort in alarm.

Signs: Tracks and droppings are similar to those of sheep and/ or deer. Goats also strip bark, but this is indistinguishable from the signs of sheep or deer. **Droppings:** Black or dark brown pellets, 1–2 cm long, cylindrical with pointed ends; more symmetrical and fibrous than sheep or deer droppings. **Tracks:** Small, two-toed prints 4·5–6·0 cm long × 3·0–4·5 cm wide. Narrow and typically well-splayed at the front, and rounded at the back, they never show the dewclaw.

Habitat: Feral herds are confined to rocky, craggy places, including hillsides, mountains and cliffs, especially above 300 m. Goats are renowned for their sure-footedness and ability to feed from precipitous slopes and craggy outcrops.

Food: Goats eat almost any vegetation, but are primarily grazers. They feed on grasses, sedges and rushes in summer, but in winter have a broader diet including gorse, tree bark, saplings and other woody material. Capable of balancing on their hind legs to reach higher up, and will even climb trees to feed. As with North Ronaldsay and Shetland sheep breeds, they eat seaweed on some islands.

Habits: Active throughout the year; diurnal, but spends parts of the day resting to digest food. Sociable, living in groups of 2–10, sometimes more; for most of the year, groups tend to be of a single sex. They hold a group home range of about 3 square kilometres.

WHERE TO LOOK/OBSERVATION TIPS

Where they occur, Feral Goats are relatively easy to locate, usually on steep rocky slopes. Good places to look include the Scottish borders, the Scottish Highlands, Snowdonia, The Burren (Co. Clare), and generally along the west coast of Ireland.

Feral Goats are very variable in colour and pattern.

Breeding behaviour: The rut is in autumn, particularly October–November. Males (bucks or billies) leave their groups to seek out fertile females (does or nannies), which they then guard from other males. Well-matched males engage in spectacular horn-clashing contests, often lasting many hours, to try and secure additional matings. 1–2 kids are born between January and March.

Population and status: Some 5,000–7,500 in Britain, but numbers unknown in Ireland. Populations vary greatly from year to year. Managed herds are often used as a conservation tool to keep trees and scrub in check, especially on precipitous sites.

Male (buck or billy) and female (nanny)

Feral Sheep
Ovis aries

Domestic sheep are familiar all over Britain and Ireland, and have been here since Neolithic times, 6,000 years ago. In a few places they live in a semi-wild state, most notably on the remote island group of St Kilda, in the Outer Hebrides. One of these breeds, known as the Soay, after one of the islands, is ancient and has strong affinities with genuinely wild sheep species in Asia. It may have been introduced to St Kilda as long as 4,000 years ago. St Kilda also hosts a second rare breed, the Boreray, introduced about 200 years ago.

Introduced (prehistoric); naturalized populations rare	
HB:	1·0–1·2 m
HS:	52–60 cm
T:	16–30 cm
Wt:	30–45 kg

Identification: Noticeably smaller and lighter-footed than domestic sheep, most Soay Sheep are a rich dark brown, often with white patches on the rump and underbelly, and sometimes on the face. Both sexes (except for about a quarter of Soay females) have horns, the males' being larger and thicker; they spiral round in typical rams' horn fashion, and have thickened rings. Boreray Sheep are also small, but tend to be creamy-white. Both breeds have short, hairy tails. Feral Goats (*page 254*) have a keel on the horns, more pointed tails, a callus on the knee and often a 'beard'.

Sounds: A bleating "*baa*".

Signs: Droppings and tracks, and tufts of wool, especially on thorns and barbed wire. **Droppings:** Small, dark and roughly cylindrical, about 1 cm across, sometimes adhering together and often left in piles. Deer droppings are usually larger or pointed at one end. **Tracks:** Two-toed print, 5–6 cm long × 4–5 cm wide. Rather rectangular in shape, with broad cleaves rounded at both ends; no dewclaws. Stride approximately 70 cm.

Soay sheep have been introduced to other locations in Britain, where flocks may be seen (see *Population and status* below for details).

Habitat: Mainly on grassland, often on steep slopes. On exposed islands, often shelters behind dry stone walls.

Food: Grazes on grass, with some Heather.

Habits: Active all year, diurnal. Lives in groups of up to 30 or more, with rams separate from the ewes and young. Each group has its own home range.

Breeding behaviour: The rut is in October–early December, when males may engage in head-butting challenges. Dominant males are promiscuous. Each ewe gives birth to 1–2 lambs in April.

Population and status: The main semi-wild population of Soay Sheep is found on the islands of Soay (where there are about 300) and Hirta (where there are about 1,200) in the St Kilda archipelago, although they have been introduced to Lundy (Devon), Holy Island (off Arran), Cardigan Island (Ceredigion) and Cheddar Gorge (Somerset). Boreray Sheep are confined to Boreray (where there are about 350–700), also in the St Kilda island group.

WHERE TO LOOK/OBSERVATION TIPS

Feral Sheep are usually easy to find where they occur. Perhaps the best way to see free-living Soay Sheep without catching a boat is to visit Cheddar Gorge in Somerset. To see Boreray Sheep, however, a visit to the island of Boreray in the St Kilda archipelago is required.

▲ *Soay sheep grazing on the island of Hirta in the St Kilda Archipelago.*
▼ *A Boreray ram*

See also: Tracks *p. 51*

Cattle

Bos taurus

Domestic cows, derived from the now-extinct wild Aurochs *Bos primigenius*, have been part of our landscape for almost 7,000 years. Some live a semi-wild existence (*e.g.* Rum, New Forest); a small herd on Swona (Orkney) has lived entirely wild since 1974, while the unique herd at Chillingham Park, Northumberland has been present for 700 years and remained unmanaged for 200 years.

Semi-domesticated	
HB:	1–2 m
HS:	1·1–1·5 m
T:	30 cm
Wt:	400–600 kg (male), 300–450 kg (female)

Identification: The many breeds of cattle differ in colour, pattern, stature, and the form of their horns. Chillingham Cattle are shaggy white, with reddish-brown ears and muzzle tip, and a dark nose; both sexes have sharp-tipped, hollow horns.

Sounds: Mooing, and a far-carrying bellow, similar to Red Deer (*page 236*) but without the roaring quality.

Signs: Droppings and tracks are easy to identify.
Droppings: The plate-sized cowpat is flat, dark brown and shiny at first, drying to a crust. **Tracks:** Two-toed prints, like those of deer or sheep, but larger: 10·0 cm long × 9·5 cm wide. The cleaves are broad, teardrop-shaped, and pointed at the front.

Habitat: Grassland, open woodland and heathland.

Food: Mainly grass, with some Heather.

Habits: Active all year and mainly diurnal; alternates feeding with ruminating, while standing or lying down. Naturally sociable and lives in herds. Most wild-living populations are all-female, or with castrated males (bullocks). At Chillingham, bulls live apart, in small groups.

This map shows the location of the semi-wild herds referred to here. However, traditional breeds of Cattle are used for conservation grazing in other locations.

Breeding behaviour: At Chillingham, high-ranking bulls are promiscuous. Females give birth to a single calf in spring or summer.

Population and status: The Chillingham herd numbers about 100.

WHERE TO LOOK/OBSERVATION TIPS

If you want to see feral cattle, they cannot be missed at Chillingham Park, where there are guided tours (chillinghamwildcattle.com).

▲ Chillingham cattle
▼ A Chillingham bull

See also: Tracks *p. 51*

Horse

Equus caballus

Wild Horses (Tarpan) were present in Britain until 9,000 years ago, and domesticated horses have been here since at least 2000 BC. Some local pony breeds are undoubtedly ancient, but despite being free-ranging in places, all populations are managed to some extent.

Semi-domesticated	
HB:	2·2–2·8 m
HS:	1·1–1·5 m
T:	15–60 cm
Wt:	150–500 kg

Identification: Horses have a bulky body, long legs with a single-toed hoof, flowing mane and wispy tail. The ancestral coat colour is dark brown, but now extremely varied. Ponies are small breeds of horse.

Sounds: Neighs and whinnies in alarm, courtship and long-distance contact; softer sounds and blows through the lips in close social contact.

Signs: Distinctive tracks and droppings; also grazes a sward very tightly. **Tracks:** Horses walk on a single toe enlarged into a hoof, in ponies 11–16 cm in diameter (depending on the size of the animal): the print is circular apart from a 'V'-shaped notch at the rear. Free-ranging horses are unshod, so the familiar horseshoe shape is lacking. **Droppings:** Brown, roughly spherical balls 3–5 cm in diameter, clearly fibrous and breaking up as they dry. Usually deposited in piles.

Habitat: Free-ranging ponies are largely found on common land, heathland, moorland and other upland habitats.

Food: Largely grazes grass, but will also browse tree leaves, gorse, Heather and moss.

Habits: Active all year and at any time of day, but often has a long sleep for part of the night. Sociable; mares with foals often associate together, and stallions keep harems if they can, becoming very active and noisy when attempting to round up straying mares. Groups live in home ranges with a core area, although several groups may come together and form larger, rather inactive gatherings, especially in summer.

Breeding behaviour: Mating takes place between May and October, peaking in June. Stallions build up a harem and compete with other males to keep it intact, baring their teeth and kicking and biting rivals. Foals are born between April and September, and suckled for about six months.

Population and status: There are several thousand free-living ponies of various breeds in Britain and Ireland, including Dartmoor, Exmoor, New Forest, Welsh Mountain, Fell, Dales, Highland, Eriskay, Shetland and Connemara. However, there is concern that some breeds are declining in numbers. Ponies are often used as a hardy conservation grazing tool.

WHERE TO LOOK/OBSERVATION TIPS

Free-living ponies are usually very easy to find, being large animals that make little attempt to conceal themselves in their heathland and moorland habitats. Welsh Mountain ponies also inhabit coastal marshes.

▲ Dartmoor pony
▼ Exmoor pony

▲ New Forest ponies
▼ Welsh Mountain pony

Red-necked Wallaby

Macropus rufogriseus

The marsupial most frequently kept in captivity in Britain and Ireland, Red-necked Wallaby originated from south-eastern Australia and Tasmania, but can do well in our climate. Several introduced colonies have persisted for many years: the biggest, of some 100 animals, being on The Curraghs, Isle of Man, which has been present since the 1970s; there is also a colony on Lambay Island, off Dublin. A previously well-established group in the Peak District has now died out. However, recent sightings have come from Cornwall, Sussex, Oxfordshire and Buckinghamshire, among other places. Other species of wallaby are kept in captivity and occasionally escape.

Introduced; rare	
HB:	65–92 cm
T:	62–86 cm
Wt:	11–27 kg

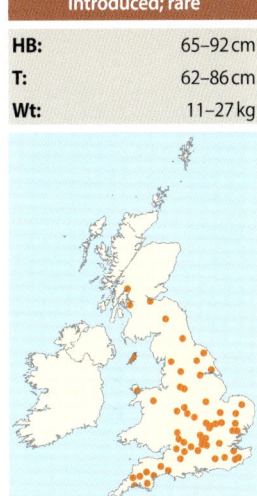

This map shows the distribution of sightings from 2008–2018 based on a study undertaken by the University of South Wales.

Identification: Small kangaroos, with an upright stance, small forepaws, long counterbalancing tail and erect ears; they move with a characteristic hop and bound on the hind legs. Red-necked Wallaby is grizzled brownish-grey with a chestnut patch on the neck and upper arm. The paws and nose are black, and the face has a whitish stripe along the upper lip.

Signs: Tracks, with long hind feet and tail drag, are unmistakable.

Habitat: Light woodland and scrub.

Food: Grazes and browses on grasses, Heather and Bracken.

Habits: Active year-round. Mainly nocturnal and crepuscular, it hides in dense undergrowth by day. Largely solitary, with a home range of about 15 ha.

Breeding behaviour: Gives birth to a single youngster (joey) in June–December, which grows in the female's pouch until emerging in May–June.

Population and status: Fewer than 150. Cold winters and traffic collisions cause significant mortality.

WHERE TO LOOK/ OBSERVATION TIPS

The Manx Wildlife Trust organises 'wallaby walks' in The Curraghs Wildlife Park, autumn/winter generally being the best time, while boat trips around Lambay Island can be booked from harbours in Dublin area.

Introduced ephemerals

With the exception of bats, a significant proportion of the terrestrial mammal species found in Britain and Ireland are not native – being either deliberate introductions, or escapes from captivity. There are, for example, only two native deer species; the other four widely established species having been introduced.

However, from time to time, other non-native mammals are found in the wild, often having been deliberately (illegally) released from captivity, or imported accidentally (such as four species of American bat). Although most of the animals concerned do not survive for very long, some are able to live 'wild' for months or even years. The term 'ephemeral introductions' has been coined for these species, and, over the years, more than 50 species have been recorded. Occasionally, escapes from captivity involve more than one individual, and small populations may become established, at least temporarily.

South American Coati

The eleven ephemeral species that are known to have bred in the wild in Britain or Ireland during the past 50 years are covered on the following pages. Most of these have, however, subsequently died out or been eradicated. At least three other ephemeral species have been recorded living in the wild for over a year, although there has been no official evidence of breeding. These are the Canadian Beaver *Castor canadensis* (see *page 98*), Crested Porcupine *Hystrix cristata* (see *page 266*), and South American Coati *Nasua nasua*, illustrated above.

See *page 319* for further information relating to non-native species.

EW Père David's Deer

Elaphurus davidianus

Introduced; not established	
HB:	1·9–2·2 m
T:	50–66 cm
Wt:	135–200 kg

Père David's Deer is extinct in its native China. In the early 20th century, the 11th Duke of Bedford gathered all 18 captive animals in the world and established a breeding population at Woburn, Bedfordshire, saving the species. There are now healthy populations in a number of safari parks from which individuals and groups sometimes wander. It favours marshes and swamps, but also found in woods and open fields, and feeds on grasses and leaves. Occurs in single-sex groups except during the breeding season from June to August. It is diurnal, and swims well.

Identification: Size of Red Deer (*page 234*) but with a short neck, narrow and elongated head, long legs and a long tail with bushy, dark tip. Antlers (males only) have almost vertical front shafts.

Black-tailed Prairie-dog

Cynomys ludovicianus

Introduced; now extinct	
HB:	36–43 cm
T:	6–11 cm
Wt:	575 g–1·49 kg

A ground-squirrel from the North American plains, which is occasionally kept in safari parks, from which it readily escapes. Feral colonies have become established over the years, some of which have persisted, including on the Isle of Wight and in Northern Ireland; currently, there are no known populations in Britain or Ireland. Naturally inhabits grassland, including grazed

areas, living colonially in burrows, sometimes at high density. Mainly eats grasses and roots.

Identification: Size of a Grey Squirrel (*page 58*), but yellowish-brown with a very short, vole-like tail with a black tip. Habitually stands on hind legs in alert posture.

Siberian Chipmunk

Tamias sibiricus

Introduced; now extinct	
HB:	12–17 cm
T:	8–15 cm
Wt:	50–120 g

Commonly kept as a pet, occasional escapes have led to the Siberian Chipmunk becoming established in parts of Europe. Several small colonies have been found in Britain, but these have quickly been eradicated and no populations are currently known. It inhabits woodland and lives in family groups in underground tunnels, where it hibernates in

winter. Diurnal and usually seen on the ground but climbs well, feeding on seeds and other plant material.

Identification: Like a miniature brown squirrel (*pages 54–61*) with five conspicuous dark brown stripes along its back and a striped, fluffy tail.

263

Coypu

Myocastor coypus

Introduced; now extinct	
HB:	37–65 cm
T:	23–45 cm
Wt:	4–10 kg

A semi-aquatic South American rodent that inhabits wetlands, living in burrows with entrances above water level. It was introduced to fur farms between 1929 and 1945 and escapees quickly established a breeding population in Britain, which eventually numbered about 200,000, most in East Anglia. It became a major pest, eating crops and causing damage by burrowing through dykes

and riverbanks. However, cold winters and an eradication campaign gradually reduced numbers, and the last individual was caught in 1989.

Identification: Like a cross between a Eurasian Beaver (*page 98*) and a Common Rat (*page 92*), with a flat forehead, large head and a long, scaly tail – but much larger than a rat.

Mongolian Gerbil

Meriones unguiculatus

Introduced; now extinct	
HB:	12 cm
T:	12 cm
Wt:	50–60 g

Gerbils are commonly kept in captivity and often escape into the wild, sometimes establishing short-lived colonies. Although recorded as far north as Yorkshire, and persisting for three years on the Isle of Wight, there are currently no known colonies in Britain and Ireland. Occurs naturally in soft soil in semi-desert habitat but introduced populations are usually near or under buildings.

Lives in family groups in burrows that can extend 60 cm underground, each group ranging over 300–1,500 square metres. Feeds on seeds and vegetation.

Identification: Larger than a mouse (*pages 84–91*), with soft yellowish-brown fur, a long hairy tail with black tip, and white underside. Various colour varieties, including black, occur in captivity.

Muskrat

Ondatra zibethicus

Introduced; now extinct	
HB:	24–40 cm
T:	19–28 cm
Wt:	0·6–1·8 kg

Native to North America, and farmed for its fur, the Muskrat was widely introduced to Europe in the 20th century and is now common across much of Western Europe. Although introduced to Britain in 1930, it was eradicated by 1937 and no populations are currently known. It inhabits lakes and rivers, living in burrows in well-vegetated banks, with entrances typically below water level. Swims well and dives frequently, and feeds on vegetation.

Identification: Rather like a pale, oversized Water Vole (*page 76*), perhaps twice as big. The tail is laterally compressed, about three times higher than wide.

VU Golden Hamster

Mesocricetus auratus

Introduced; now extinct	
HB:	12–16·5 cm
T:	13–15 cm
Wt:	80–150 g

Golden Hamsters are very common in captivity and have occasionally escaped and established small feral populations, invariably in urban and suburban areas with outhouses, sheds, *etc*. Although a population persisted in North London for a few years in the 1980s, this has since died out; there are currently no feral populations known in Britain or Ireland. This rodent is native to Syria and Turkey, where it inhabits steppe grassland; it is mainly nocturnal and lives in burrows up to 2 m deep. Feeds mainly on grain, which it stores in cheek pouches, but will also eat insects.

Identification: Larger than a Bank Vole (*page 69*), smaller than a rat (*pages 92–97*): soft-furred, blunt-faced with a very short tail. Most have whitish fur on the underparts and flanks, but very variable.

Himalayan Porcupine

Hystrix brachyura

(Malayan Porcupine, Himalayan Crestless Porcupine)

Introduced; not established	
HB:	63·0–72·5 cm
T:	6–11 cm
Wt:	up to 12·5 kg

In the 1970s, a small population of Himalayan Porcupine became established in Devon, persisting for over a decade. Despite occasional reports of sightings, there are currently no known populations. It usually lives in small family groups in woodland and is nocturnal, spending the day in a den underground.

Identification: Unmistakable: very large with a black front half covered with bristles and the back half with black-and-white quills, which it rattles menacingly when threatened. It lacks large spines on top of the head shown by the similar Crested Porcupine *Hystrix cristata*, also recorded wild.

Striped Skunk

Mephitis mephitis

Introduced; not established	
HB:	57–80 cm
T:	18–39 cm
Wt:	1·2–6·3 kg

Native to North America, the Striped Skunk is famous for its ability to spray a foul-smelling substance up to 4 m towards an intruder. Surprisingly, it is often kept as a pet, but with the scent glands removed. However, since 2007 it has been illegal to remove the scent glands, and it is likely that individuals were released into the wild as a result: a small population became established in the Forest of Dean, which may still persist.

It is mainly solitary, living in a burrow with several entrances, and is an opportunistic omnivore, active mostly at night or at twilight throughout much of the year.

Identification: Unmistakable, sleek and boldly patterned, about the size of a domestic cat, with short legs and a long, bushy tail. Black-and-white pattern variable: some largely black with 'V'-shaped stripes; others white with black legs.

Raccoon

Procyon lotor

Introduced; not established	
HB:	60–95 cm
T:	19–40 cm
Wt:	4·0–15·8 kg

Native to North America. Escaped individuals have often been recorded in Britain, some remaining at large for up to four years. Although breeding has been recorded on at least one occasion, there are currently no known established populations. It is largely nocturnal and solitary, and occurs in a wide range of habitats, including urban areas, and is strongly associated with water.

If introductions continue, there is a risk that this species may become established and invasive, as it has in mainland Europe.

Identification: Unmistakable cat-sized omnivore with dark 'bandit'-style facemask and black-and-white ringed tail.

VU Asian Short-clawed Otter

Aonyx cinerea

Introduced; now extinct	
HB:	36–46·8 cm
T:	22·5–27·5 cm
Wt:	2·4–3·8 kg

Commonly kept in wildlife parks and private collections, from which it has sometimes escaped and bred in the wild in Britain. There were a number of reports from around Oxford from 1983–1993, and occasional sightings elsewhere, but none are believed to be in the wild in Britain or Ireland at present. In its native range in south-east Asia and China it inhabits lakes

and rivers, feeding mainly on crabs, but also fish and small mammals. It is diurnal and particularly active at dawn and dusk.

Identification: Looks like a miniature Otter (*page 210*), with similar coloration and a snake-like tail, but has a short, blunt face and incomplete webbing on the toes.

Marine Mammals

For more detailed information, see the companion WILD*Guides* title *Britain's Sea Mammals*

Marine mammals are not a distinct biological grouping, but are defined instead as mammals that have a reliance upon the marine environment for feeding – although not necessarily for breeding. Hence, both **pinnipeds** (seals and Walrus) and **cetaceans** (whales, dolphins and porpoises) are considered marine mammals – pinnipeds being dependent upon the sea for food, but dry land for resting, moulting and breeding, and cetaceans being wholly dependent upon the marine environment throughout their life-cycle. The Otter (*page 210*) is not considered to be a marine mammal as it is not strictly dependent upon the marine environment, being primarily a riverine species. The Otters that live in coastal areas of, for example, Shetland, need to bathe in fresh water in order to rid themselves of sea salt that would otherwise reduce the insulative qualities of their fur.

Pinnipeds

All seals (and Walrus) belong to a subgroup of marine mammals collectively known as pinnipeds. They are classified as part of the order Carnivora (see *page 17*), sharing common ancestors with, amongst others, bears and dogs – an interesting parallel given the shared curiosity Grey Seals and domestic dogs often show towards one another.

All pinnipeds share some basic features and, while generally cumbersome and ungainly on land, are well adapted to the marine environment. The limbs have evolved into webbed flippers, and the body is streamlined for manoeuvrability underwater – even their reproductive organs are retractable into the main body mass to minimise drag when swimming. They also have a thick layer of blubber that provides insulation, and a circulatory system that redirects blood away from the body surface to minimise heat loss. Further adaptations are: nostrils that close when submerged; eyes with a protective clear membrane that allows excellent vision both above and below the water; and whiskers which, in conditions of low visibility, provide information about their surroundings. When diving, pinnipeds regulate the flow of blood to their organs, and are able to empty their lungs completely without ill-effect when diving deeply.

Pinnipeds are carnivorous and hunt by pursuing their prey. They feed on a wide variety of marine fishes, cephalopods and molluscs, although they may feed opportunistically on seabirds and even other seals. Although most pinnipeds are generalist feeders, some species are specialists, such as the Ringed Seal that feeds almost exclusively on crustaceans.

Cetaceans (see also *pages 280–281*)

Whales, dolphins and porpoises belong to the order Cetacea of marine mammals, collectively known as cetaceans. Like terrestrial mammals, cetaceans are warm-blooded, breathe air into their lungs through their nostrils, and suckle their young. However, unlike any other mammals (except for the Dugong and the manatees) cetaceans live exclusively in water.

Fish (shark)

Cetacean (dolphin)

The adaption of cetaceans to life in the marine environment has been so profound that many have evolved to resemble fish (sharks in particular) in both form and structure. The orientation of the tail should be noted when comparing sharks and cetaceans; in the former, the tail is orientated vertically, and in the latter the flukes are in a horizontal plane.

All cetaceans breathe air through blowholes situated on or near the top of the head and they have a greater tolerance to carbon dioxide than any terrestrial mammal, with lungs that are way more efficient at processing oxygen; their ancestral front limbs have evolved to form flippers, and their hind limbs have vanished altogether to create an efficient hydrodynamic body shape. Like pinnipeds, they have a layer of fat, known as blubber, beneath the skin to insulate them.

Cetaceans come in all shapes and sizes, from the huge 24–30 m-long filter-feeding Blue Whale to the compact 1·5 m-long fish-hunting Harbour Porpoise, and have evolved to be able to exist in all manner of aquatic habitats. There are two suborders of cetacean (see *page 17*):

Baleen whales (Mysticeti) employ a filter-feeding system that uses plates of keratin edged with bristles located along the upper jaw in a close arrangement. When a whale opens its mouth, water pours in and (in the rorquals) the ventral pleats expand to accommodate the water. The whale then pushes its tongue forward (and contracts its pleats) to force the water out through the baleen plates, the bristles trapping any prey inside the mouth.

Toothed whales (Odontoceti) have teeth on the upper and lower jaw – which is often elongated to form a distinct 'beak'. These species tend to be smaller than the baleen whales and hunt by pursuit. Many toothed whales find their prey in deep and dark or murky waters using echolocation. A high frequency click is produced by passing air through a special bony 'nostril' situated near the blowhole. This click is directed and magnified through a fatty organ located in the whale's 'melon' (the often rounded or bulbous forehead). Any returning echo is picked up by another fatty structure located in the lower jaw and is transmitted to the ear, enabling the animal to pinpoint its prey.

Common Seal

Phoca vitulina

(Harbour Seal)

Despite its name, this species is neither as numerous as the Grey Seal, nor as widespread. However, where it does occur it is easy to see, and it will often haul up on beaches and sandbanks in tidal estuaries.

Identification: Smaller and slimmer than Grey Seal (*page 272*), but the two species can be surprisingly tricky to tell apart (see *page 274*), both showing great variation in their basic colour pattern, from almost black through rusty brown to off-white. Common Seal tends to have little contrast between back and belly and is densely, more uniformly speckled, rather than blotched. It has 'V'-shaped nostrils and a more cat-like face, with a more concave profile and stepped forehead. The most reliable identification feature is the position of the eye when viewing the animal's head in profile: the eyes of Common Seal are a third of the way from the tip of the muzzle to the back of the head (midway on Grey Seal). Males and females look very much alike. Pups have a similar coloration to adults, unlike young Grey Seal pups.

Sounds: Quiet on land, uttering a few growls, and pups wail at mothers. Less vocal than Grey Seal.

Signs: Tracks of seals are distinctive in mud or sand, a series of depressions either side of a dragged body trail. Imagine the track a rowing boat with two oars would make.

Habitat: The characteristic seal of sand flats and tidal estuaries, but also occurs in sea lochs and on low islets and along some rocky coastlines. Feeds out to sea in relatively shallow water.

Food: Feeds on locally abundant fish of various kinds, including sand-eel, herring, sprat and cod, plus squid, octopus and some crustaceans.

Rare seals (pp. 275-279)
LEGALLY PROTECTED
Biodiversity List (En, Sc, NI)
Locally common

HB:	1·3–1·9 m
	(pup 65–100 cm)
Wt:	60–150 kg
	(pup 8–12 kg)

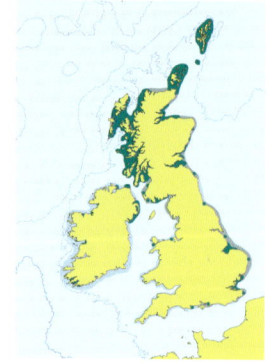

Fur can range from dark (BELOW) to very pale (RIGHT), but is nearly always evenly speckled or blotched.

WHERE TO LOOK/OBSERVATION TIPS

Found mainly around the coastline of Scotland and eastern England. Can be seen all year round hauled out on beaches and sandbanks, such as Blakeney Point (Norfolk) and Donna Nook (Lincolnshire), as well as many reliable sites in Scotland. Commercial seal-watching boat trips operate in many places, such as Blakeney.

Habits: Active all year and at any time of day, alternating between periods of solitary foraging and communal resting, when they can be found on land at a favoured haul-out spot, especially at low tide. Feeding usually takes place near to the haul-out, although seals have been recorded making return trips of over 100 km. Common Seals can be aggressive at haul-outs, and as a result individuals may be more widely spaced apart than Grey Seals (which often share the same haul-out). Usually shy of people and boats. In August they gather to moult.

Pups are born on sandbanks and take to the water soon after, with the rising tide.

Breeding behaviour: Breeding season is June–July (Grey Seal Sep–Dec). Common Seals are not strictly colonial. Mating occurs in July, with a single pup being born (typically on a tidal sandbank or rocky skerry) the following June–July. There is an initial (but crucial) bonding process and after a few hours the pup is able to take to the water in the close company of its mother. The pup is suckled for 3–6 weeks, with pups more than doubling their body weight during this period. As soon as a pup is weaned, the mother will mate again. Mating generally takes place in water, and some males mate with several females.

Population and status: Not as numerous as Grey Seal, with about 46,000 individuals at moult period in Britain, mostly in Scotland; about 5,000 in Ireland. Populations fluctuate, and there have been recent declines in Orkney and Shetland. From 1998–2002 an outbreak of Phocine Distemper Virus reduced the population, especially on the east coast. As well as being covered by EU and domestic legislation, in England Common Seals are managed under the Conservation of Seals Act 1970, and in Scotland under the The Marine (Scotland) Act 2010.

Will haul out on rocks or on beaches.

Grey Seal

Halichoerus grypus

The Grey Seal is a fairly common sight around the coasts of Britain and Ireland, so it might come as a surprise to know that it is one of the world's rarest seals. Britain and Ireland hosts about 36% of the world population.

Identification: Larger than Common Seal (*page 270*) but with a similar wide variation in coat colour and pattern, and it can need care to tell the two species apart (see *page 274*). However, typical individuals are readily separable. Grey Seals are dark grey above, with an obvious contrast to the pale cream underside, and with large, very irregular and sometimes well-spaced dark blotches. They have almost parallel nostrils and a long, flattish muzzle giving a 'Roman nose' or horse-like shape. Male Grey Seals have a much longer muzzle than females, which are closer in profile to Common Seal. The most reliable identification feature is the position of the eye when viewing the animal in profile: the eyes of Grey Seal are situated midway between the end of the muzzle and the back of the head (one-third on Common Seal). Males are usually darker than females, with youngsters the palest of all. Pups are fluffy and a creamy yellowish-white for the first 3–4 weeks of their lives during which they grow and moult a new sleek coat similar in appearance to that of a pale adult female.

Sounds: Sometimes noisy, making a moaning sound with melancholy, lost air, plus hisses and growls at the colony.

Signs: As Common Seal.

Habitat: Spends much of its time in the sea, in the shallower waters above the Continental Shelf. However, even in the non-breeding season it spends up to 40% of its time hauled up on beaches, sandbars and islands, evidently to rest after foraging. During the breeding season it spends much time ashore, usually on remote islands, rocky beaches and in sea caves where there is not much disturbance. They may occur up to 300 m from the sea in large colonies.

Often inquisitive, and can appear remarkably dog-like when in the water.

◉ Rare seals (*pp. 275-279*)	
LEGALLY PROTECTED	
Locally common	
HB:	1·8–2·1 m
	(pup 95–105 cm)
Wt:	105–310 kg
	(pup) 11–20 kg

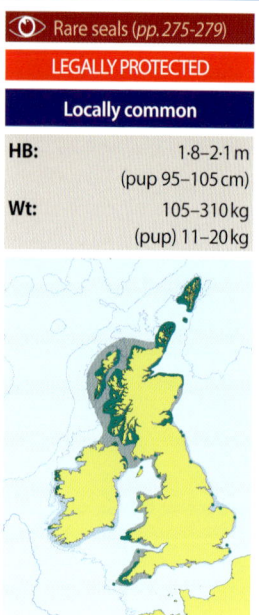

WHERE TO LOOK/OBSERVATION TIPS

Easy to see all year and especially numerous in Scotland. Donna Nook (Lincolnshire) is the most accessible English site – the animals are particularly approachable at pupping time (November–December) but care must be taken to avoid disturbance. Other popular sites include Skomer (Pembs.) and Blakeney Point (Norfolk); the Irish stronghold is the Inishkea island group, Co. Mayo.

Food: Mainly fish, including sand-eel, cod, plaice, haddock and herring, as well as squid. Occasionally takes swimming seabirds.

Habits: Active all year, and at any time of day; often rests at low tide. Divides its time between solitary seagoing foraging trips with sociable hauling out on beaches, islands, rocks or sand-bars. Typically, leaves a haul-out site for a 2–5 day foraging trip before returning or transferring to another haul-out, some animals wandering widely outside the breeding season. Feeds up to

For the first month pups are covered in creamy-white fur which is then moulted to a sleek adult-like coat.

50 km from a haul-out. Once in the water it spends most of its time beneath the surface, with dives to over 300 m depth that usually last up to 10 minutes. The number of animals at a haul-out ranges from a few to thousands. Individuals do not exhibit much aggression, but there is often sexual and age segregation. At breeding colonies females keep about 3 m apart with aggressive displays.

Breeding behaviour: The breeding cycle is annual: mating occurs during September–December, (the exact time varying by location), with females giving birth the following season. The single pup is born on land, fed on milk for only about 18 days and then abandoned, after which the female is ready to mate again. They are colonial, with males and females returning to the same favoured site each year. Males compete for groups of females in parts of the colony, and will mate with several females if they can; they repel other males with aggressive calls and gestures, but fights are few.

Population and status: The UK population is 182,000 and now stable after a period of increase. The Irish population is 5,500–7,000. Grey Seals are vulnerable to overfishing, pollution and marine waste, and disturbance at their colonies. They are covered by EU and domestic legislation and managed under the Conservation of Seals Act 1970, but can be shot under licence in Scotland.

Will haul out on rocks or on beaches, when its blotched coat is most apparent.

Seals compared

To identify Common Seal and Grey Seal focus on details of the head.

COMMON SEAL

GREY SEAL

Common Seal nostrils are diagonal, forming a small, neat 'V'-shape; those of the Grey Seal are vertically aligned parallel slits when closed.

COMMON SEAL

GREY SEAL ♀

GREY SEAL ♂

Common Seal has a much more concave or 'dished' face with a shorter, neater muzzle – they look 'friendlier', with eyes set one third of the way back from the muzzle to the back of the head.

Grey Seal typically has a convex, or at least flat, long muzzle, giving a doleful, 'Roman-nosed' appearance; eyes are halfway between the tip of the muzzle and the back of the head.

Common Seal (LEFT) has a coat which is more uniformly speckled and has less contrast between back and belly compared to the more blotched contrasting coat of the Grey Seal (RIGHT).

Walrus and rare seals

On occasion, Walrus and four rare seal species make their way from their usual home in the Arctic to Britain and Ireland. Understandably, the vast majority of sightings are from Shetland, but there have records all down the east coast of Britain as well as in the south west.

VU Walrus   *Odobenus rosmarus*

Many people would be surprised to know that this magnificent Arctic animal has occurred in British and Irish waters, but there have been over twenty records since 1900, .

Identification: Instantly recognisable: no other seal-like animal has tusks. The tusks are whitish and highly apparent, although they are absent in juveniles and are often broken and occasionally lacking in adults. Also easily identified by its wrinkled skin, large size, bull-necked bulk, short, stiff whiskers (vibrissae) set on a moustache-like pad, and by its oddly square-shaped head with small bloodshot eyes and nostrils set atop the muzzle.

Habitat: Sometimes sighted far out to sea (follows pack-ice in winter), but equally may haul out on rocky or sandy shores.

Food: The principal diet is bivalve molluscs gathered from the sea bed, clams being the favourite. Sometimes catches and eats seabirds, other seals and even cetaceans, and will scavenge carcasses.

Habits: Walruses are fond of spending time hauled out on land to rest and moult. In the water, typically surfaces every five minutes but a dive of 24 minutes has been recorded. Although generally sociable, those seen in Britain have been single wanderers. Uses tusks in social interactions, and also as an aid for hauling out on land.

Population and status: Vagrant, particularly Shetland and the west coast of Ireland (most recent: one seen first in Orkney then at various places along the coast of mainland Scotland, the Outer Hebrides and Shetland (March–June 2018); one Ireland and Pembrokeshire (March 2021)).

LEGALLY PROTECTED	
Vagrant	
HB:	up to 3·6 m (male), up to 3 m (female)
Wt:	up to 1,900 kg (male), up to 1,200 kg (female)

VU Hooded Seal

Cystophora cristata

A large and fairly distinctive rare seal from Arctic waters. Adult males have a fleshy inflatable 'hood', although almost all records from Britain and Ireland are of females and immatures.

Identification: Likely to be noticed as a large seal with a marbled pattern; superficially similar to a female Grey Seal (*page 272*) but has more contrasting blotches. Both sexes are essentially silvery whitish-grey with a variegated pattern of dark blotches and black around the muzzle. Females have smaller blotches and are whiter, so the blackish head is more contrasting. The head is large and broad and, in males, there is an obvious fleshy 'trunk' that droops down over the mouth. Juveniles are blue-grey on the back, dark on the head and flippers, with contrasting white below (to 14 months of age).

Habitat: Arctic ice-floes, and the open sea. In Britain it has been found on beaches and islands, often among other seals.

Food: Fish, molluscs (octopus) and crustaceans. A deep diver, submerging frequently to 100–600 m for 5–25 minutes.

Habits: Generally solitary outside the breeding season and vagrant records, unsurprisingly, have involved single animals. Rather shy. Has particularly heavy and awkward movements on land. Individuals may wander far from their usual range; extralimital animals have been found as far south as Portugal.

Population and status: Vagrant: more than 20 records since 1900 (most recent: Co. Cork, Ireland (January 2020)). Although not specifically protected by domestic legislation, the Hooded Seal is covered by EU legislation.

Vagrant	
HB:	2·5–3·0 m (male),
	2·0–2·4 m (female)
Wt:	300–400 kg (male)
	160–230 kg (female)

Bearded Seal

Erignathus barbatus

The most frequently recorded of the rare seals, with visiting individuals often staying for prolonged periods. A large and powerful species, the Bearded Seal probably spends more time wandering in the open sea than other Arctic seals, which may explain its more regular occurrence here.

Vagrant	
HB:	2·1–2·7 m
Wt:	200–430 kg

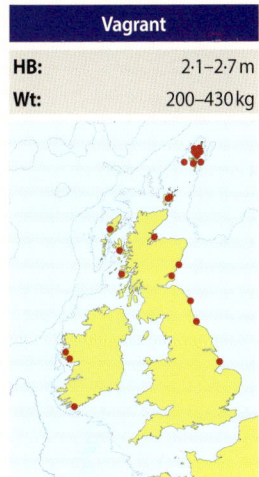

Identification: A large, plain-coloured, grey or brown seal marked with few, if any, blotches on the body. Named for its impressive set of long, bushy white whiskers, some of which curl inwards at the tip when dry. In profile the head looks far too small for the long, heavy body, even for a seal. The short front flippers are square-ended, with all five digits the same length, a unique feature. There are often pale patches around the eyes and on the muzzle.

Habitat: An animal of the shallow waters around the Arctic pack-ice. In Britain usually seen hauled out on beaches or islands.

Food: It has a broad diet, encompassing crabs, shrimps, molluscs, fish (especially cod and flatfish) and even some marine algae. Animals forage on the seabed and dives are often down to 45 m and sometimes much more.

Habits: A rather solitary and relatively shy animal. May adopt a vertical position in water in times of danger.

Population and status: Vagrant: more than 25 records since 1900, and virtually annual in recent years. Although not specifically protected by domestic legislation, the Bearded Seal is covered by EU legislation.

Harp Seal

Phoca groenlandica

A handsome seal of the High Arctic pack-ice, extremely rare in Britain but unlikely to be overlooked because of its striking pied pattern.

Vagrant	
HB:	1·6–1·9 m
Wt:	140–190 kg

Identification: Adults unmistakable with a dark hood and broad, almost black stripes running from the shoulder to the pelvis, which look vaguely like harps, hence the name. The stripes meet at the shoulders and form a 'V'-shape when seen from above. Males have bold black markings that contrast with the otherwise silky whitish fur; females are a little more subdued, with slightly less bold blotches against a greyer ground colour. Juveniles/immatures have silvery greyish fur with scattered small dark spots. The Harp Seal is medium-sized but with noticeably small flippers, whitish whiskers and a broad, round head.

Habitat: Fundamentally an Arctic animal, associated with pack-ice. The vagrant individuals found in Britain have been on rocky or sandy beaches and islands.

Food: Fish and crustaceans (krill).

Habits: Very sociable, gathering into thousands to breed and moult; however, strays to Britain are usually alone. Hauls out on land/rocks and often adopts 'bottle' upright posture in water.

Population and status: Vagrant: 14 live records since 1900, but just three since 2000 (most recent: Outer Hebrides (February 2016)). In addition, there are two 2018 records (not shown on map) of young animals washed up (from Isle of Skye (May) and Aberdeen (July)). Although not specifically protected by domestic legislation, the Harp Seal is covered by EU legislation.

Ringed Seal

Pusa (Phoca) hispida

A wide-ranging species – the commonest seal in the Arctic and an annual winter visitor to northern Iceland. There is also a population in the Baltic, where the animals are much larger in size, as well as freshwater populations in Finland (Lake Saimaa – endangered) and Russia (Lake Ladoga). It very rarely strays to Britain and Ireland although, given its abundance in the Arctic and its similarity to Common Seal, it may be overlooked.

Vagrant	
HB:	1·1–1·5 m (Arctic)
	1·5–1·7 m (Baltic)
Wt:	50–70 kg (Arctic)
	110–124 kg (Baltic)

Identification: Distinguished from the Common Seal (*page 270*) is the pattern of whitish rings on the dark back, making the skin look as though it is peeling. However, beware of the highly variable Common Seal, which can show a similar pattern, although with less clear rings. On Ringed Seal the underside is usually plain light grey; on Common Seal there are usually marks on the underside. More subtle is the difference in shape: Ringed Seal is small but very stout and plump (the girth is 80% of the length) and it has a very short neck, a small head and a stubby muzzle, making the eyes look large.

Habitat: In Britain, will haul-out on beaches and islands in similar habitat to other seals.

Food: Mainly small fish (especially Arctic Cod) in summer, with a broader diet including crustaceans in other seasons.

Habits: Breeds on pack-ice or in snowdrifts, where it makes a lair (snow tunnel) for breeding or resting. Moves south in autumn as the pack-ice builds. Rather shy.

Population and status: Vagrant: More than ten records since 1900 (most recent: Co. Kerry (same individual also Co. Clare), Ireland (January 2020)). Although not specifically protected by domestic legislation, the Ringed Seal is covered by EU legislation.

Identifying cetaceans

Naming the parts

Cetaceans have specially adapted bodies. The technical terms used in this book are shown on the annotated illustrations here.

Upper jaw
Rostrum
Blowhole (double)[1]
Dorsal fin
Tail-stock
Lower jaw
Tail flukes
Ventral pleats[2]
Pectoral flipper
Notch
Melon
Blowhole (single)[1]
Beak

Humpback Whale

Northern Bottlenose Whale

[1] Baleen whales have a double blowhole; toothed whales have a single blowhole

[2] Only the baleen whales of the family Balaenopetridae have ventral pleats, and are collectively know as rorquals

What to look for

Cetaceans can be difficult to encounter, and a challenge to identify. They are often distant, may only give brief views before diving and usually all you see is the head, back and dorsal fin. Some key points to help identification are below.

Dorsal fin: the location of the fin on the back and the shape/size of the dorsal fin.

Blow (see *opposite*)**:** the plume of moisture-laden air expelled as the animal surfaces and exhales is often visible from a considerable distance, and is sometimes distinctive. All cetaceans blow when they surface, but only those of the larger whales are usually visible.

Head shape: details of the head (if seen!) are very useful (for example, Risso's and Bottlenose Dolphins can look similar, but are readily identified by the blunt head of the former and the beak of the latter).

Size: size can be difficult to estimate, especially as cetaceans usually do not show much of their body and it is often hard to judge distance and scale – try to compare what you see with a nearby object such as a boat, buoy or passing seabird.

Dive sequence (roll): the order in which the blow, head, fin and tail appear can be useful for identification. Dolphins have short, fast rolls, whereas whales have more regular longer-lasting rolls.

Colour pattern: although many cetaceans are plain-coloured, some, especially the dolphins, are distinctive and unmistakable.

Short-beaked Common Dolphins reveal their beak and distinctive colour pattern when porpoising.

Surface behaviour: what the animal does on the surface is often helpful in identification. Some behaviours to look for are:

- **Porpoising:** moving rapidly through the water by alternating leaps out of the water with fast swimming just below the surface (see photo *opposite*).
- **Spy-hopping:** raising the head vertically out of the water (see *page 290*).
- **Logging:** resting motionless at the water surface (see *page 301*).

RISSO'S DOLPHIN

MINKE WHALE

- **Tail-slapping:** the tail is lifted clear of the water and slapped down on the surface.
- **Breaching:** jumping out of the water wholly (full breach) or partially (half-breach).

RISSO'S DOLPHIN

- **Chorus line:** a group of animals surfacing together in a line.

BOTTLENOSE DOLPHIN

- **Bow-riding:** riding the pressure wave created ahead of a moving vessel (or whale!).

Blows compared

NORTH ATLANTIC RIGHT WHALE	HUMPBACK WHALE	FIN WHALE
'V'-shaped: Bowhead Whale, North Atlantic Right Whale	**Mushroom-shaped:** Sperm Whale (angled left), Humpback Whale	**Columnar** (small to large): Minke Whale, Sei Whale, Fin Whale, Blue Whale

Minke Whale

Balaenoptera acutorostrata

This is by far the commonest whale around British and Irish coasts, the only one that regularly occurs in shallow water and close to land. Hotspots include the west coast of Scotland (especially the Western Isles) and the coast of north-east England. It occurs year round, with a peak May–September.

Identification: Small for a whale but still much longer and larger than any dolphin. Rather sleek, with a long body and pointed head. The rostrum is bisected by a single ridge that runs from the front of the blowholes to the tip. Has a distinctive sickle-shaped dorsal fin situated nearly two-thirds of the way along the back. Appears black at a distance, but a close view shows it to be dark grey above, with grey flanks that fade to white on the belly. A diagnostic feature is a white band across the upper surface of the flipper, although this can usually be seen only at close range (see breaching – *page 281*). Northern Bottlenose Whale (*page 302*) can show a similar back and fin profile, but has a beak, distinctive head shape and brownish coloration.

Habitat: Shallow water over the continental shelf, including estuaries.

Food: Small fish (*e.g.* sand-eel, herring, sprat) and squid.

Habits: Found all year, but more common from May–September and peaking in abundance in late summer. It is not very sociable, usually encountered as individuals or mothers with calves. Often seen in the company of flocks of Gannets, Manx Shearwaters and Kittiwakes.

Breeding behaviour: Mating and calving occur December–June. The single calf is born after 10–11 months, then fed on milk for 4–6 months. Individuals are sexually mature at 3–8 years.

Population and status: UK population 10,500, trend unknown; unknown numbers off Ireland. Threatened by pollution, by-catch in trawlers and the vestiges of commercial whaling.

👁	Northern Bottlenose Whale (*p. 302*) Sei Whale (*p. 300*)

LEGALLY PROTECTED

Biodiversity List (En, Wa, Sc, NI)

Shallow water resident

Length:	7–10 m
Weight:	5–10 t

DIRECTION OF TRAVEL

The inconspicuous vertical blow is likely to seen only if the whale is close, and/or in calm conditions.

single ridge on rostrum

WHERE TO LOOK/OBSERVATION TIPS

Can be tricky to observe at sea, because of its habit of rolling only once or twice before diving and disappearing. This is the only whale likely to be seen from land in Britain and Ireland. During the summer and early autumn an evening walk in calm weather along a cliff-top on the west coast of Scotland may provide a sighting. Commercial whale-watching trips, such as those operating off western Scotland and Co. Cork in Ireland offer the best chance of an encounter.

Blow: light, low (less than 2 m), vertical and bushy; often not visible at all. **Dive sequence:** surfacing roll is quite fast; typically, there are 5–8 surface rolls/blows at approximately one minute intervals, then a longer dive of about five minutes. **1]** the inconspicuous blow appears at the same time as the head and snout, which emerge at an angle; **2]** next to show are the top of the head and the back (NB sometimes the head and back appear with the blow); **3]** next the dorsal fin appears; **4]** the animal then rolls quickly, exposing the tail-stock, which sinks beneath the surface without the tail flukes showing. **Behaviour:** often breaches, especially during rough weather. Quite inquisitive and may spy-hop around boats.

▲ *The position of the dorsal fin, two-thirds of the way along the back, becomes apparent mid-roll.*
▼ *Typical view of a Minke Whale as it first breaks the surface.*

VU Fin Whale

Balaenoptera physalus

This impressive whale is the second largest animal on the planet, behind Blue Whale. Although regular in British waters, the Fin Whale is not easy to see and a prolonged boat or ferry trip into suitable territory provides the best chance of an encounter. In Ireland it can be seen regularly from the south and south-west coasts almost year round.

Identification: A very long but relatively slim and streamlined whale identified by its huge size, substantial blow, relatively small 'swept-back' dorsal fin (located three-quarters of the way along the back) and distinctive profile on deep diving. The left lower jaw is dark and the right lower jaw is white – an asymmetry unique to Fin Whale. The back is dark grey, sometimes with subdued white chevron marks behind the blowhole; the flanks are grey; the throat and belly are white. The rare Sei Whale (*page 300*) is similar in form but smaller and with a proportionally larger, upright sickle-shaped dorsal fin located two-thirds along the back. The very rare Blue Whale (*page 304*) is larger, mottled grey-blue and with a much smaller, distinctive dorsal fin.

Habitat: Mainly deep waters off the continental shelf, but in summer regularly reaches the 500 m depth contour.

Food: Diet mostly small crustaceans, but also larger crustaceans (krill), small squid, and fish such as herring, mackerel, sand-eel and whiting obtained mainly by lunge-feeding.

Habits: Typically encountered in ones and twos although, several may gather at very rich feeding grounds. Capable of swimming at 45 km/h and cruising at 30 km/h. Dolphins and seabirds, such as Gannets, are often seen associating with feeding Fin Whales.

Breeding behaviour: Mates October–January, first breeding when 6–12 years old. One calf is born in warmer waters, and is weaned at 6–7 months.

Sei Whale (*p. 300*) Blue Whale (*p. 304*)	
LEGALLY PROTECTED	
Biodiversity List (En, Wa, Sc)	
Deep water resident	
Length:	18–26 m
Weight:	40–80 t

DIRECTION OF TRAVEL ⟶

The tall, columnar blow is visible at a great distance and can take several seconds to dissipate.

← DIRECTION OF TRAVEL

WHERE TO LOOK/OBSERVATION TIPS

Usually seen May–August (peak August) in Britain, and May–January (peak October–January) off southern Ireland. Try a boat/ferry trip in the Minches, St George's Channel or from Co. Cork.

Blow: a dense, tall column 4–8 m high, much more conspicuous than Minke (*page 282*) or Sei Whale blows; usually 2–5 blows at approximately 5–10 second intervals before a deep dive of 10 minutes. **Dive sequence: 1]** the blow and usually just the very top of the head appears first; **2]** the first part of the back shows; **3]** then the rest of the back dorsal fin; **4]** the animal then rolls forward without showing the tail. **1–4]** is repeated until prior to a deep dive when; **5]** the tail-stock is characteristically heavily arched, and submerges after the dorsal fin. **Behaviour:** only occasionally breaches, but does so in style, creating a huge splash.

Population and status: Regular but sparse summer visitor to Britain, with some records at other times. Sighting patterns suggest that almost certainly there is a small resident population in the waters south and west of Ireland that spends the summer offshore, and the winter inshore.

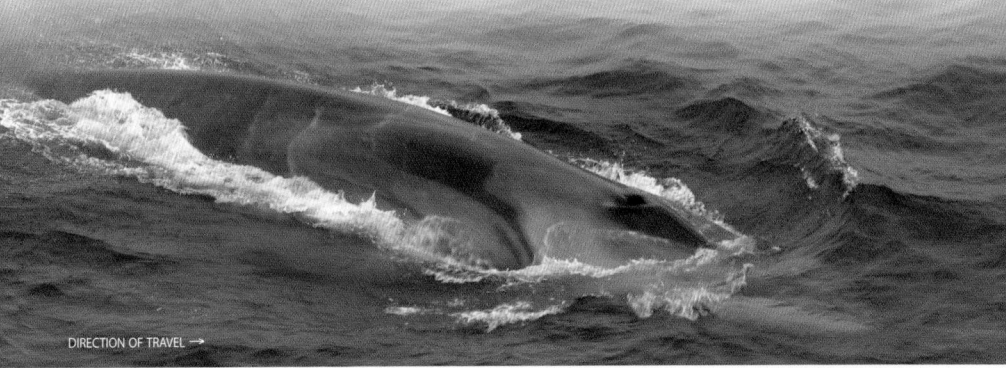

DIRECTION OF TRAVEL →

▲ *Has distinctive flank chevrons and unique white right jaw.*
▼ *Typical view of a Fin Whale as it first breaks the surface and* (INSET) *the classic profile as it begins a deep dive.*

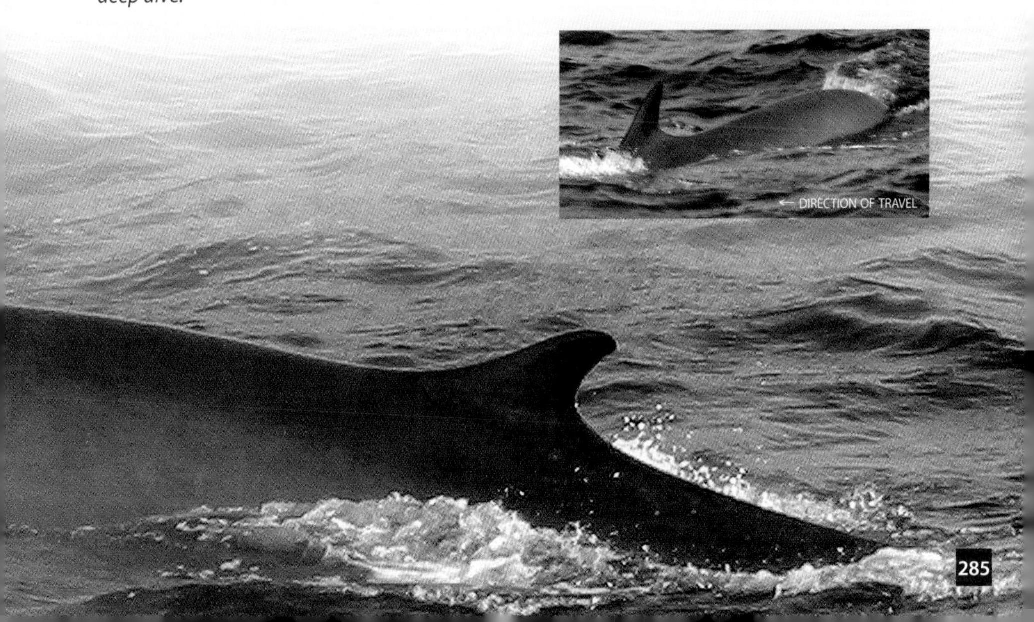
← DIRECTION OF TRAVEL

Humpback Whale

Megaptera novaeangliae

Highly demonstrative and distinctive whale that has become a regular visitor to the region in recent years.

Identification: A large, portly whale most readily identified by its unique long, wing-like, mainly white flippers. Black or dark grey on the back and flanks, with a white belly. It has a broad, flattened head covered in fleshy, knobbly tubercules. The dorsal fin, situated two-thirds of the way along the back, is usually short and stubby. The tail flukes have a very characteristic serrated edge, and a pattern of black and white on the underside that is unique to the individual. Although similar in size to Sperm Whale (*page 301*) and the very rare North Atlantic Right and Bowhead Whales (*page 303*), Humpback Whales are readily identified by their distinctive flippers, dive sequence and fluke pattern.

Habitat: Mainly deep waters, but sometimes seen from land.

Food: Small fish (sand-eel, herring, anchovy and mackerel), krill (crustaceans) and plankton. Often lunges into shoals of fish. Groups of Humpback Whales will co-operate when feeding, synchronising their feeding lunges or trapping prey in a 'bubble net', by swimming in decreasing circles below large shoals and blowing bubbles to trap the fish inside.

Habits: Migratory; present in our waters from about June to February. Encountered as singletons or single figure parties. Irish sightings appear to be linked to movements of their herring and sprat prey.

Population and status: A regular and increasing visitor, now showing signs of recovery from the catastrophic impact of past persecution by commercial whalers.

← DIRECTION OF TRAVEL

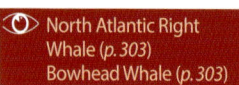

👁 North Atlantic Right Whale (*p. 303*)
Bowhead Whale (*p. 303*)

LEGALLY PROTECTED
Biodiversity List (Wa, Sc, NI)

Migrant

Length:	11–15 m
Weight:	25–40 t

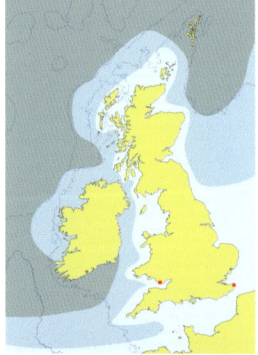

← DIRECTION OF TRAVEL

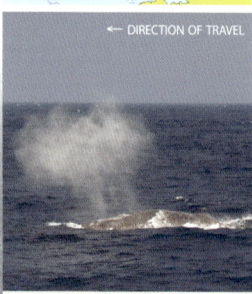

Has a distinctive mushroom-shaped blow.

Typical low profile of a Humpback Whale breaking surface at the start of a roll.

WHERE TO LOOK/OBSERVATION TIPS

Best looked for in three areas during the summer: the Northern Isles south to eastern Scotland; the northern Irish Sea north to south-west Scotland; and the Celtic Sea. Sumburgh Head (Shetland) is a well-known site. Sightings in Ireland occur from late July to early February, wth a peak in November.

Blow: usually tall (up to 3 m), vertical and mushroom-shaped, but variable.
Dive sequence: 1] the blow and head appear first; **2]** followed by the first part of the back; **3]** the back arches high and the whale sinks. **1–3]** is repeated until **4]** prior to a deep dive, the back and tail-base are arched more steeply; and finally **5]** the tail flukes break the surface. Dives typically last 3–15 minutes (average 6 minutes). **Behaviour:** sometimes extremely demonstrative at the surface; breaching frequently often landing on its back, sometimes lifting the whole body clear of the water, but more often half breaches; also flipper-slaps and tail-slaps; frequently spy-hops and lob-tails.

The long white flippers are unique.

▲ *The tail flukes have a diagnostic serrated rear edge and a variable white pattern.*
▼ *Mid-roll, shows a distinct hump in front of the short, stubby dorsal fin.*

DD # Killer Whale (or Orca) *Orcinus orca*

The big and boldly patterned Killer Whale is the largest dolphin in the world. Although rare, it visits Shetland and other parts of northern Scotland annually.

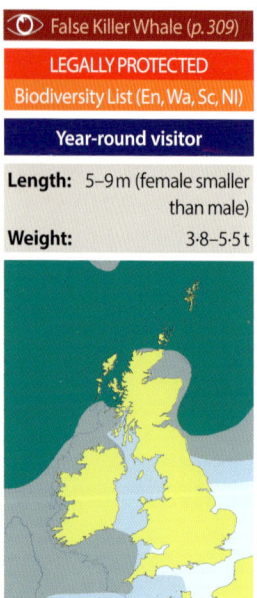

👁 False Killer Whale (*p. 309*)
LEGALLY PROTECTED
Biodiversity List (En, Wa, Sc, NI)
Year-round visitor

Length: 5–9 m (female smaller than male)
Weight: 3·8–5·5 t

Identification: The Killer Whale, one of the 'blackfish', can hardly be confused with anything else given its bulk, striking body pattern and prominent, centrally located dorsal fin, which differs according to age and sex; adult males have broad-based sharply triangular fins up to 2 m high; females and young have more sickle-shaped fins, which can appear similar to Bottlenose Dolphin (*page 292*) at a distance. The big head has a blunt beak. The body is mainly black with a grey saddle behind the dorsal fin. The throat and belly are white and there is a distinctive white patch above the eye. The dorsal fin is located halfway along the back. The flippers are big and rounded. The rare False Killer Whale (*page 309*) is mainly black.

Habitat: Coastal or deep waters, dependent on the group type.

Food: Offshore populations hunt fish such as herring and mackerel, whereas inshore groups have a broader diet encompassing seals and seabirds. Killer Whales also attack other cetaceans.

Habits: Occurs year-round, but most commonly seen May–September. Found for the most part in stable family groups of 2–30 animals; coastal populations in smaller pods. Groups usually contain both sexes and at least one adult male. Sometimes curious around boats.

Breeding behaviour: Calving October–January, gestation 15–18 months. The single calf is nursed for at least one year. Individuals breed at age 15 (males), 8–10 years (females).

Population and status: The world's most widespread cetacean (with potentially several distinct species involved), uncommon in British and Irish waters. No population estimates.

Females (LEFT) have a sickle-shaped dorsal fin, whereas those of males (RIGHT) are tall and triangular.

← DIRECTION OF TRAVEL

WHERE TO LOOK/OBSERVATION TIPS

The most reliable place and time is Shetland in June and July, where with patience and good local contacts you stand a reasonable chance of success. Sightings are also regular in various parts of north and west Scotland, as well as far offshore, associated with mackerel purse-seine fishing boats during November–March.

Blow: relatively tall and bushy. **Dive sequence: 1]** the blow, head and dorsal fin appear at the same time; **2]** then the first part of the back followed by the prominent dorsal fin; **3]** then the tail-stock is arched prior to diving. Performs a dozen or so short dives before a longer one. **Behaviour:** often breaches, leaping almost vertically upwards, spy-hops, tail-slaps and flipper-slaps.

▲ Often seen in groups with individuals spy-hopping (as here), breaching and tail-slapping.
▼ A powerful male travelling is a majestic sight, the white eyepatch and grey saddle are usually visible.

Long-finned Pilot Whale

Globicephala melas

Long-finned Pilot Whale is the most regular 'blackfish' dolphin in the region. Highly sociable animals with strong, stable social units, perhaps best known for their unfortunate habit of mass-stranding, when tens or even hundreds may beach together.

Identification: Pilot whales are obviously larger than most dolphins, and easily identified by the unusual forward position of the dorsal fin (situated just one third of the way along the back). They have a stocky body, a bulbous, rounded head and long flippers. The dorsal fin shape differs with age and sex: adult males have an especially broad-based, lobe-like fin that curves strongly backwards; those of adult females are less curved; those of juveniles are pointed and dolphin-like.

Habitat: Mainly deep waters, with a strong preference for the continental shelf slope at the 1,000 m depth contour. Sometimes seen in shallower waters.

Food: Cuttlefish and other small squid, plus some shoaling fish. Usually feeds at night and rests by day.

Habits: Present all year, but easiest to see in July–August. Usually found in family groups of 2–50. Not shy of vessels.

Breeding behaviour: Calves in April–September. The single offspring is born after 12–16 months gestation, then looked after by the female and is weaned at 18–44 months. Individual cows give birth every 3–6 years. Unusually for cetaceans, both male and female calves remain in their mother's pod.

Population and status: Resident and fairly common, but population unknown. Affected by by-catch in trawler nets, whaling (particularly infamously in the Faroe Islands) and pollution.

False Killer Whale (*p. 309*)
Melon-headed Whale (*p. 309*)

LEGALLY PROTECTED
Biodiversity List (En, Wa, Sc, NI)

Deep water resident

Length:	3·5–6·5 m
Weight:	up to 3 t

Frequently seen spy-hopping, and also exhibiting other behaviours.

A mature male making fast progress.

WHERE TO LOOK/OBSERVATION TIPS

Most common in northern and western Scotland and south-west England, where a search from a coastal watchpoint from late summer to autumn could bear fruit. Otherwise a dedicated trip to deeper water might be required.

Blow: low and distinctly bushy, often not visible. **Dive sequence: 1]** the distinctive, bulbous head appears first; **2]** then the dorsal fin; **3]** followed by the long back and tail-stock (revealing the forward position of the dorsal fin); **4]** then the long back and tail-stock are arched before diving (with no sign of the flukes). Long dives are made, on average of 10 minutes duration. **Behaviour:** in daylight often seen logging; breaching at various angles regular, but for some reason in British waters this is not often seen; much more likely to be seen spy-hopping, tail-slapping or flipper-slapping.

▲ A female/immature showing the characteristic deep dive profile with the forward position of the dorsal fin and strong tail stock on view.
▼ A typical family group: the mature male, told by its distinctive broad-based fin, is in the centre.

Bottlenose Dolphin   *Tursiops truncatus*

(Common Bottlenose Dolphin)

The Bottlenose Dolphin is undoubtedly the easiest cetacean to see in Britain and Ireland, as there are several places where it can reliably be expected to be present. Although Bottlenose Dolphins can easily be seen from shore, local boat companies often run dedicated trips which can provide a better view.

Identification: A large, robust, somewhat plain dolphin. The most conspicuous feature is the tall, sickle-shaped dorsal fin, which is centrally positioned and sharply pointed. The curved forehead leads into a short, stubby beak (the 'bottlenose'). The grey back fades to slightly paler flanks and a lighter belly, lacking the strong patterns of other dolphins.

Habitat: Occurs both inshore and offshore, as separate populations. Inshore groups are often drawn to estuaries, headlands and sandbanks where there are strong tidal currents.

Food: Highly varied, including fish (*e.g.* cod, herring, sprats, sand-eels), squid and shellfish. Feeds at various depths (offshore animals diving down to 500 m) and uses echolocation, sight and normal hearing to detect prey.

LEGALLY PROTECTED
Biodiversity List (En, Wa, Sc, NI)

Inshore and offshore resident	
Length:	1·9–3·9 m
Weight:	150–650 kg

Habits: Active throughout the year and probably at all times of day and night. Several inshore populations are present all year and are highly sedentary, offshore groups move around. Extremely sociable, and typically encountered in pods of up to 50 animals, occasionally more, but sometimes encountered singly. Females with calves may form close-knit groups, and sometimes members will 'babysit' calves while their colleagues go foraging. Individuals frequently co-operate when feeding, attempting to herd fish into shallower water or towards rocks. They can be quite aggressive, and will frequently skirmish among themselves. Occasionally, they may attack and kill other cetaceans, especially Harbour Porpoises (*page 298*), which are competitors for food. Bottlenose Dolphins sometimes associate with other cetaceans, such as Long-finned Pilot Whales (*page 290*).

A porpoising individual will reveal the beak and plain flanks that confirm identification.

WHERE TO LOOK/OBSERVATION TIPS

Easy to see throughout the year at certain strongholds: the Moray Firth (including the justly celebrated Chanonry Point, best at high tide) and Cardigan Bay. Other good sites include the Shannon Estuary, south-west Scotland, the outer reaches of Aberdeen harbour (worth watching from the ferry to and from Shetland), southern and western Ireland, Cornwall and off the Dorset coast.

Dive sequence: on occasion all you may see is the sickle-shaped fin breaking the surface, with the beak barely visible and with no sign of the tail. However, it may travel with conspicuous porpoising and big splashes. **Behaviour:** extremely active. Routinely jumps clear out of the surface, making a large splash; it also commonly bow-rides vessels of all kinds, or rides the wake instead. May even wake-ride larger whales. It also lob-tails and head-slaps – this is often a very demonstrative animal.

Breeding behaviour: May breed at any time of year although births peak May–September. Males compete for access to females, often aggressively. Females give birth on a two-year cycle. The single calf, born a year or so after mating, is dependent on milk for up to 20 months, and may continue to suckle for some years and remain with their mother for many years, even after the mother gives birth again. Females are mature at 5–13 years old; males at 9–14.

Population and status: Several hundred are resident in British and Irish waters, and others occur as migrant visitors. Its strongholds in the Moray Firth and Cardigan Bay are protected under the EU Habitats Directive (see *page 312*). Dolphins are threatened by the depletion of fish stocks, pollution, collisions with craft and excessive noise disturbance.

▲ *A longish, straight back in front of a large dorsal fin is indicative of Bottlenose Dolphin.*
▼ *Inshore, groups can often just be seen moving slowly in a favoured area.*

Short-beaked Common Dolphin   *Delphinus delphis*

(Common Dolphin)

Highly energetic and sociable dolphin, usually seen in groups, – a large pod at sea can be a truly dramatic spectacle, with animals creating sizeable splashes as they leap and porpoise.

Identification: Medium-sized beaked dolphin most easily identified by its colour pattern. The back is dark grey, forming a distinctive point where it runs into the central part of a tawny-yellow and grey 'figure-of-eight' flank pattern. The belly is whitish. Two black stripes run from the beak; one to the flipper and one to the eye. The centrally located dorsal fin is rather tall and triangular with a backward pointing tip, often with a central pale patch.

Habitat: Prefers deep waters outside the 200 m contour, but is sometimes seen close inshore.

Food: Fish such as sprat, mackerel, pilchard and whiting.

Habits: Present all year, although some groups evidently move into our waters May–October. Highly sociable, in pods of 20 on average, but can be 500 strong. They often associate with other cetaceans, principally Striped Dolphins (*page 299*), but occasionally with Bottlenose Dolphins (*page 292*) and Long-finned Pilot Whales (*page 290*). They are very fast, routinely travelling at 60 km/h.

Breeding behaviour: Breeds June–September, with a single calf born 10–11 months later, after which the female mates again. The calf feeds on its mother's milk for 11 months, but can take solid food from about three months of age.

Population and status: Numerous, particularly well offshore. The continental shelf population is estimated at 63,000.

Dive sequence: usually seen porpoising at speed, progressing by intermittent leaps out of the water. In this species the porpoising leaps are low, causing quite small splashes. **Behaviour:** very active and fast-moving, routinely leaping clean out of the water (but less acrobatic than Striped Dolphin; regularly bow-rides.

LEGALLY PROTECTED	
Biodiversity List (En, Wa, Sc, NI)	
Resident and migrant	
Length:	1·7–2·5 m
Weight:	70–110 kg

WHERE TO LOOK/ OBSERVATION TIPS

Quite common well out to sea in the western English Channel, the southern Irish Sea, the Sea of Hebrides and the Minches, and off southern Ireland; and sometimes farther north. Although may be seen from western headlands, summer ferry crossings and whale-watching trips offer much better chances of an encounter.

no branching flank streak (compare with Striped Dolphin)

grey on flank and tail stock

ochre and grey 'figure-of-eight' pattern

Atlantic White-sided Dolphin

Lagenorhynchus acutus

A powerful and acrobatic dolphin of northern waters. It occurs off the extreme northern and western coasts of Britain and Ireland, and, in summer, in the northern North Sea. It sometimes mixes in pods with White-beaked Dolphins.

Identification: Like a bigger, chunkier, more powerful and not so playfully exuberant Short-beaked Common Dolphin. Best identified by the sharply demarcated white patch on the flanks, which extends along the tail stock as an elongated yellow or tan stripe (in Short-beaked Common Dolphin the ochre panel is at the front). Otherwise black above, grey in the middle and white on the belly. It is also distinguished from Short-beaked Common Dolphin by the thick tail stock and short, stubby beak (a feature shared with White-beaked Dolphin (*page 296*), which lacks sharply defined flank streaks).

Habitat: Essentially cool waters along the continental shelf edge (100–500 m depth), but with seasonal visits to shallower waters.

Food: Wide variety of fish (including mackerel, whiting and lantern fish) and some squid.

Habits: Present all year in deep water north of Scotland, visiting inshore waters June–September. Very sociable, and usually seen in pods of 2–15 (up to 50), with pods sometimes coalescing. It is a powerful and swift swimmer at the surface, often attaining speeds of 14 km/h without porpoising. It does not tend to approach boats.

Breeding behaviour: Breeds May–August, with a single calf born 11 months later, after which the females mate again. The calf is weaned at 18 months.

Population and status: Population around Britain and Ireland not known, but not numerous and much commoner to the north.

LEGALLY PROTECTED	
Biodiversity List (En, Wa, Sc)	
Resident; **inshore summer visitor**	
Length:	2·0–2·8 m
Weight:	165–230 kg

WHERE TO LOOK/ OBSERVATION TIPS

Difficult to see in Britain and Ireland, although summer ferry trips to the Northern and Western Isles, or expeditions to the west of Orkney, offer a reasonable chance of an at-sea encounter.

Dive sequence: when travelling, all that is usually seen is the upper back and dorsal fin.
Behaviour: very active, given to frequent breaching, lob-tailing (the rear end held vertically above the water) and tail-slapping. It only occasionally bow-rides.

elongated yellow or tan stripe

White-beaked Dolphin

Lagenorhynchus albirostris

The White-beaked Dolphin is restricted to the North Atlantic and, although not well known, is quite common in British and Irish waters.

Identification: A large, stocky dolphin with a very short white-tipped beak. The black, centrally-located dorsal fin is tall and hooked – similar to other dolphins. However, White-beaked Dolphin has a sooty black back with a distinctive white or pale grey 'saddle' behind the dorsal fin and two softly defined grey areas on the flanks (one ahead of the dorsal fin and one behind) – unlike the sharply defined panels of Atlantic White-sided Dolphin (*page 295*) or Short-beaked Common Dolphin (*page 294*). The belly is white.

Habitat: Shallow water, less than 200 m deep (usually 50–100 m).

Food: Various fish, including mackerel, herring, sand-eel and flatfish; also squid and octopus.

Habits: Present all year, but sightings peak June–September. It is almost always seen in small groups of up to 20, and usually less than 10. These often hunt co-operatively, herding and panicking their prey. Sometimes mixes with other species, including Atlantic White-sided Dolphins. When swimming fast can reach speeds of 30 km/h.

Breeding behaviour: The single calf is born after 10–11 months gestation, in May–August, with a few into October. Individuals are sexually mature at 7–12 years.

Population and status: Estimate of 20,000 in British and nearby waters.

LEGALLY PROTECTED	
Biodiversity List (En, Wa, Sc)	
Resident	
Length:	2–3 m
Weight:	165–230 kg

WHERE TO LOOK/ OBSERVATION TIPS

Localized: hotspots include the central and northern North Sea and Western Isles (peak August); uncommon in Ireland. Can be seen from land in Shetland (Sumburgh Head) and Aberdeen (peak August) and the Northumberland coast (peak May–June). Pelagic trips run from north-east England and into Lyme Bay, Dorset (July–August).

Individual beak coloration varies from white to pale grey, some with dark markings.

Dive sequence: when travelling at speed, some of the head may break the surface, followed by a typical dolphin roll exposing the back and then the tail stock. **Behaviour:** very active, with breaching, lob-tailing and tail-slapping. Also bow-rides boats.

pale 'saddle' just behind dorsal fin

Risso's Dolphin   *Grampus griseus*

This uncommon large dolphin is found mainly in the west, from the Southwest Approaches to the Northern Isles.

Identification: A large, robust dolphin with a distinctive pale grey body that is often covered with whitish scratches and scars. The bulbous head has a small melon and lacks a beak. The centrally-located dorsal fin is tall, erect and slightly hooked, and it has long, pointed flippers. Young Risso's Dolphins are dark and unscarred, and may be confused with Bottlenose Dolphin (*page 292*), but the obvious beak of that species should eliminate confusion. Long-finned Pilot Whale (*page 290*) is black and the dorsal fin is located much farther forward.

Habitat: Primarily continental shelf waters 200–1,000 m deep.

Food: Octopus, cuttlefish, squid and krill, taken mainly at night.

Habits: Present all year, but commonest May–October. It may be found singly, but usually occurs in groups of up to 15, sometimes more. Can be aggressive – the scratches on the bodies of older individuals are a result of skirmishes. Sometimes associates with Long-finned Pilot Whales.

Breeding behaviour: Resident pods calve December–July. The single youngster is born after 13–14 months gestation.

Population and status: Population unknown, but not numerous.

LEGALLY PROTECTED
Biodiversity List (En, Wa, Sc, NI)

Resident

Length:	2·8–3·8 m
Weight:	300–500 kg

WHERE TO LOOK/ OBSERVATION TIPS

Easiest to see in August and September, with the Northern and Western Isles being a hotspot, most often from vessels well offshore. They may also be seen from coastal watchpoints here and in Devon, Cornwall, the Lleyn Peninsula and from headlands of western Ireland.

Young Risso's Dolphins are dark and unscarred; reminiscent of Bottlenose Dolphin.

Dive sequence: slow moving animals surface in a leisurely fashion: **1]** first the head followed by; **2]** the back and dorsal fin; **3]** then the flat back and fin before; **4]** rolling over slowly exposing the dorsal fin and tail-stock (and sometimes the flukes). **Behaviour:** does not porpoise like other dolphins; it can be demonstrative (breaching, spy-hopping and tail-slapping), but is often seen just logging or slowly swimming in a 'chorus line' (see *page 281*).

blunt headed

mature animals very pale, usually with scars

Harbour Porpoise

Phocoena phocoena

Harbour Porpoise is the most numerous inshore cetacean in Britain and Ireland, but its small size and shy, lethargic nature mean that it is easily overlooked, even in calm conditions.

Identification: The smallest cetacean found in British and Irish waters. Although inconspicuous, it is quite easy to recognise by its diminutive size and small, triangular, blunt-tipped dorsal fin, situated centrally along the back. The head is small and rounded, and lacks a beak. The upperparts are dark grey, fading to lighter grey on the flanks, with a white underside. Young animals often have brownish backs. Dolphins are much larger, more lively and have much taller, sickle-shaped dorsal fins.

Habitat: Shallow water over the continental shelf, with a preference for coastal inshore waters.

Food: Shoaling fish, such as herring, sprat, whiting and sand-eel.

Habits: Present all year and can be seen from almost any coast, at any time. Markedly shy and often avoids vessels. Found in small, loose groups (1–6), or just mother and calf. It usually feeds alone, but occasionally groups of animals may feed co-operatively.

Breeding behaviour: Mating takes place April–September, with peak July–August. A single calf is born the following May–August. Calves remain with their mother and can feed themselves within 2–3 months, but will still take milk for up to 11 months.

Population and status: Continental shelf population 386,000; it has suffered a long-term decline and is the subject of conservation measures. However, it is still fairly common, especially in north-west and north-east Scotland, south-west England, Wales and the western and southern coasts of Ireland. Threats include by-catch in fishing nets, pollution and declining fish stocks.

LEGALLY PROTECTED
Biodiversity List (En, Wa, Sc, NI)

Inshore resident	
Length:	1·5–1·7 m
Weight:	50–70 kg

WHERE TO LOOK/ OBSERVATION TIPS

Look from land, using binoculars, when the sea is flat-calm. July–August are the best months; hotspots include the Northern Isles, the west coast of Scotland, Wales and southern Ireland. December–March in SW England can also be good.

Dive sequence: 1] the top of the head and back as far back as the dorsal fin all appear together in a very shallow roll; **2]** the head then submerges, with the back and fin still visible; **3]** finally the rest of the animal submerges with the fin being the last part to remain above the surface. **Behaviour:** slow and lethargic; hardly ever breaches; does not bow-ride. Sometimes moves at speed at surface, producing a 'rooster tail' of spray, but more often just logs on the surface.

Readily identified by its small, short and triangular dorsal fin halfway along the back.

← DIRECTION OF TRAVEL

Striped Dolphin

Stenella coeruleoalba

A dolphin of warm temperate and tropical seas, regluarly seen in the Bay of Biscay, but rare around Britain and Ireland.

Identification: Very similar in build and demeanour to Short-beaked Common Dolphin (*page 294*) and the two are very hard to tell apart at a distance unless the flank pattern is seen: Striped Dolphin has a pale grey flank streak that goes from the eye and **branches up towards the dorsal fin**. A narrow black stripe runs from the beak, through the eye and forms a neat border between the grey flanks and white/pinkish belly, although this is only seen if an individual jumps clear of the surface. Another black stripe runs from the eye down to the flippers.

Habitat: Deep water far offshore, beyond the continental shelf.

Food: Fish such as sprat, anchovy and whiting, plus some small squid and crustaceans.

Habits: Usually seen July–September, as singles or schools up to 30, sometimes mixed in with Short-beaked Common Dolphins. Tends to be shy of boats, and groups will keep tight together and move slowly near them. Often seen accompanying Fin Whales (*page 284*), one of the best ways to spot them.

Population and status: A rare visitor to British and Irish waters, but may become more frequent with rising sea temperatures.

LEGALLY PROTECTED	
Biodiversity List (Wa, Sc)	
Offshore resident	
Length:	1·8–2·5 m
Weight:	90–156 kg

WHERE TO LOOK/ OBSERVATION TIPS

The best chance is to take a late summer pelagic trip out into the Western Approaches.

Dive sequence: typically porpoises at speed.
Behaviour: even more acrobatic than Short-beaked Common Dolphin, with breaching, leaping (as high as 7 m) and somersaulting all routine. A closely packed pod with lots of splashing and leaps of varying heights is likely to be this species.

Often leaps clear of the surface, showing the striped flank pattern.

EN Sei Whale

Balaenoptera borealis

Very difficult to see in British and Irish waters, being generally rare and unpredictable. It is mainly a deep water species that only occasionally wanders within sight of land.

Identification: Very long and slender grey whale similar in form and colour to a small Fin Whale (*page 284*). The main identification feature is the distinctive dorsal fin located two-thirds the way along the back. The fin is upright and sickle-shaped, and appears proportionately larger than that of the similar Fin and Minke Whales (*page 282*). Other key differences are that Sei Whale has a much larger blow than Minke Whale, and has a dark right jaw (white in Fin Whale).

Habitat: Mainly deep waters outside continental shelf.

Food: Zooplankton, small fish and squid.

Habits: A summer visitor, mainly July–October. Usually seen singly or in pairs. Tends to be wary of boats. It has a habit of feeding on plankton just below the surface by skimming, unlike other rorquals. During their dives, which last 5–20 minutes, they rarely descend more than a few metres below the surface and, in calm conditions, can be easily tracked by their flukeprints. When feeding or travelling the back and dorsal fin are visible for a longer period than Fin or Minke Whales.

Population and status: Rare but regular visitor, numbers fluctuate year on year.

LEGALLY PROTECTED	
Biodiversity List (En, Sc, NI)	
Offshore summer visitor	
Length:	12–21 m
Weight:	15–30 t

WHERE TO LOOK/ OBSERVATION TIPS

A boat trip to the west of Shetland in summer may produce a sighting. Often feeds at dawn.

Blow: tall, thin and vertical, not as robust as Fin Whale but more prominent than Minke Whale.
Dive sequence: 1] when surfacing, the blow and fin usually appear at more or less the same time (unlike most Fin Whales, but similar to Minke Whale); **2]** the animal then rolls slowly at a shallow angle with the dorsal fin prominent; **3]** then usually sinks (rather than rolls) with the dorsal fin the last part visible. The tail stock is not usually visible during the dive sequence (unlike Fin Whale).
Behaviour: does not breach often, and then only at a low angle, belly-flopping back into the water.

Sei Whale can look very like Minke Whale mid-roll (*page 283*), but the dorsal fin is taller and more upright. If there is any doubt, wait because …

… if it sinks with the dorsal fin last to submerge, it is a Sei Whale.

The head and dorsal fin usually appear together, but beware confusion with young Fin Whale.

Sperm Whale
Physeter macrocephalus

The Sperm Whale hunts large squid in pitch-black depths down to 3,000 m. The head contains an organ that produces a jelly-like substance, spermaceti, that may help in echolocation.

Identification: The largest of the toothed whales, some individuals approaching the size of a Fin Whale (*page 284*). Its unusual profile, which often looks like a log floating in the water, is due to the flat-topped head which constitutes almost one third of its total length. There is a small, triangular hump in place of the dorsal fin, and often a series of smaller humps or 'knuckles' down to the tail. The wrinkled skin is normally slate-grey or brown. The tail flukes are triangular, generally smooth-edged and have a deep central notch.

Habitat: Deep water, especially over canyons and the slopes of the continental shelf.

Food: Principally large and medium-sized squid, including Giant Squid (*Architeuthis*); and occasionally fish, including sharks.

Habits: In summer, found in deeper water, but some appear to overwinter (December–February) in shallow water off north-west Scotland. Usually seen in groups of up to 20, comprising immature males and females with calves. Adult and adolescent males travel separately and return to the females for the breeding season.

Population and status: Uncommon visitor, population unknown. The North Atlantic population may be increasing after decades of persecution finally ended in the 20th century.

LEGALLY PROTECTED
Biodiversity List (En, Sc)

Mainly winter visitor

Length:	11–18 m (male larger than female)
Weight:	13·5–55·8 t

WHERE TO LOOK/ OBSERVATION TIPS

Best chance is a dedicated trip into deep water off north-west Scotland.

Blow: angled forwards to the left; typically 1–2 m high, every 10–30 seconds for 5–8 minutes, until the animal deep dives. **Dive sequence: 1]** the distinctive angled blow appears first; **2]** then the head, back and dorsal hump slowly emerge; it may remain in this profile for some time, either motionless or travelling slowly, until it dives when; **3]** it rolls forward in a steep arch revealing the dorsal hump before; **4]** the tail-stock rises and the animal lifts its flukes vertically as it dives. **Behaviour:** spends a long time on the surface, preceding a dive that can last up to two hours; full or half breaches are frequent, often performed in bad weather and by juveniles. It also frequently spy-hops, lob-tails and flipper-slaps.

Distinctive blow, angled to the left; this whale is travelling away and right.

← DIRECTION OF TRAVEL

The steep arch and dorsal hump are normally a precursor to a deep dive…

… when the tail is revealed prior to a near vertical descent.

DD **Northern Bottlenose Whale** *Hyperoodon ampullatus*

A scarce and little-known cetacean which hunts squid in deep water during dives that can last well over an hour. One famously ended up in the River Thames in 2006.

Identification: One of the 'beaked whales', the largest of the group seen in Britain. It is quite distinctive with an enormous swollen melon above the stubby protruding beak. It is typically dark brown to dark grey above, with older males showing a lot of 'powdery' white on the melon and beak. The dorsal fin, set two-thirds of the way along the back, is sickle-shaped and quite tall, similar to that of Minke Whale (*page 282*), although head shape and colour should eliminate confusion.

Habitat: Shows a strong preference for water more than 1,000 m deep, often at the edge of the continental shelf. Only rarely found in shallow water, and then perhaps only if sick or disorientated.

Food: Primarily deep-water squid.

Habits: Present in British and Irish waters March–October, peaking July–September. At sea encountered in pods of up to ten animals, which are markedly inquisitive and may approach stationary boats. They often log on the surface, blowing every 30 seconds or so.

Population and status: Scarce but regular summer and autumn visitor. Sightings are increasing.

Blow: low and bushy, just 1–2 m high. Slightly angled forwards.
Dive sequence: 1] the distinctive bulbous melon appears first, sometimes at a steep enough angle to reveal the beak, with the blow visible in calm conditions; **2]** the back and dorsal fin appear, with the top of the head usually still visible; **3]** then the animal rolls languidly before **4]** submerging with the tail-stock slightly arched (the flukes do not show). **Behaviour:** this whale makes long, deep dives, so it often logs motionless on the surface for 10 minutes or more. It only occasionally breaches, leaping well clear of the water.

LEGALLY PROTECTED	
Biodiversity List (Wa, Sc)	
Offshore summer visitor	
Length:	7–9 m
Weight:	5·8–7·5 t

WHERE TO LOOK/ OBSERVATION TIPS

Try taking a ferry or a dedicated whale-watching trip off northern Scotland in July–September.

The melon and short beak are diagnostic of beaked whales in the region.

Large whales without a dorsal fin and with a 'V'-shaped blow

Both the North Atlantic Right Whale and the Bowhead Whale are large whales that lack a dorsal fin and have a 'V'-shaped blow (if seen from the front or rear). Once common, they were hunted to near extinction. Recent sightings here, and elsewhere in the Atlantic, may reflect a slow population recovery by both species following the cessation of commercial whaling.

CR North Atlantic Right Whale LP BL (En, Sc) *Eubalaena glacialis*

no hump in front of blowhole   callosities

DIRECTION OF TRAVEL →

Length:	10·0–15·6 m
Weight	12·1–20·3 t

Only a small population of this whale remains, ranging off the Atlantic coast of Canada and the USA. The most recent confirmed sighting in our area was off Co. Donegal in 2001.

Identification: North Atlantic Right Whale has diagnostic white callosities on the head and jawline and lacks a pronounced bump in front of the blowhole. The chin is black and the underside has a variable amount of white. The flukes are very large, smooth-edged, with pointed tips and a deep central notch. **Habits:** Seen alone or in small groups. Rather tame and approachable. **Status:** Very rare, although once frequent in British and Irish waters.

Blow: distinctive, high, 'V'-shaped if seen front/rear on. **Dive sequence: 1]** the distinctively calloused head and the blow appear together first; **2]** followed by the back (with the top of the head and often part of the jawline still visible); **3]** the animal rolls and the lack of dorsal fin becomes apparent. **1–3]** is repeated unless the animal deep dives, in which case **4]** the flukes appear. It tends to remain 2–3 minutes on the surface at a time. **Behaviour:** often breaches and is quite demonstrative.

Bowhead Whale LP BL (En) *Balaena mysticetus*

pronounced hump in front of blowhole

DIRECTION OF TRAVEL →

Length:	up to 20 m
Weight	up to 90 t

A rare whale, normally found along the edge of the Arctic pack ice. Two recent sightings off south-west England.

Identification: Bowhead Whale has a pronounced bump (splashguard) in front of the blowhole and lacks any callosities. The underside is black and usually there is a white chin patch with a row of black spots along the edge. **Habits:** Generally solitary, although mother and calf pairings occur. **Status:** Three records of juveniles – Isles of Scilly (Feb 2015), Mount's Bay, Cornwall and Carlingford Lough, Ireland (both May 2016). The 2016 records are of the same individual, and may also account for the 2015 record.

Blow: distinctive, high, 'V'-shaped if seen front/rear on. **Dive sequence: 1]** the head (with the distinctive blowhole hump) and the blow appear first; **2]** followed by the back; **3]** the animal rolls and the lack of dorsal fin becomes apparent. **1–3]** is repeated unless the animal deep dives, in which case **4]** the flukes appear.

EN **Blue Whale** LP BL (Sc)

Balaenoptera musculus

The largest animal that has ever lived, only occasionally seen in British waters. A krill-eating species of colder, open ocean waters, it sometimes occurs close to the continental shelf. Hydrophone (underwater microphone) studies suggest that it may occur regularly in deep water to the north and west of Britain and Ireland, although seeing one of these giants in the region is an enormous challenge.

Identification: The huge blow is usually the first indication that a Blue Whale is nearby. If the animal breaks surface its enormous size, coloration and behaviour are unmistakable. However, it usually lies just below the surface of the water, with little showing, but even so the pale greyish, mottled coloration can often be seen. When travelling, or preparing for a dive, the body roll is prolonged, during which the diagnostic tiny dorsal fin, just over two-thirds of the way along the back is usually revealed. The very large, long head has an obvious bump (splashguard) on top. **Habits:** Usually encountered alone or in pairs. **Status:** On the basis of hydrophone evidence, it may be more abundant (though this abundance is relative) in British waters in November–December but very rarely seen in British and Irish waters, or indeed in the north-east Atlantic.

| Length: | 24–30 m |
| Weight | 80–150 t |

Blow: massive, dense column up to 10 m high, every 10–20 seconds.

Dive sequence: 1] the huge blow appears first; **2]** slowly, the first part of the back (being such a large animal, the dorsal fin is still below the surface); **3]** the dorsal fin appears as the whale continues to roll smoothly; **4]** eventually revealing most of the tail stock, but not the tail. **1–4]** is repeated unless the animal is doing a deep dive, in which case **5]** the flukes appear at a shallow angle.

Behaviour: young animals occasionally breach, albeit somewhat half-heartedly.

▲ *Has an obvious splashguard in front of the blowhole* (LEFT); *the huge tail flukes are revealed when an animal commences a deep dive* (RIGHT).
▼ *During the long body roll, the diagnostic tiny dorsal fin is usually seen.*

Fraser's Dolphin LP

Lagenodolphis hosei

Length:	2–2·6 m
Weight	160–210 kg

Blow: insignifcant.
Behaviour: quite exuberant, leaps and splashes and can swim fast. It frequently breaches and sometimes falls backwards into the water.

Usually found in deep tropical waters. There is one record of a stranded individual from the Outer Hebrides in 1996.

Identification: Distinctive bold markings and an unusual shape. A very small, stubby beak; the flippers and dorsal fin are also very small, the latter set halfway along the back, and triangular. The body is stocky, and mostly lead-grey on the upperside. A dark (often black) stripe from the beak widens into a mask around the eye and continues down the flank. There is another black line from the eye to the flipper. The whitish underside is sometimes flushed pink. **Habits:** Usually in large pods and has a tendency to mix with other species. **Status:** Vagrant, one record.

Cuvier's Beaked Whale LP BL (En, Wa, Sc)

Ziphius cavirostris

Length:	5–7 m
Weight	2·0–3·4 t

Blow: low (<1 m) and bushy, slightly angled forwards; 2–3 at 20–30 second intervals.
Dive sequence: 1] head and melon appear first (sometimes at an angle that reveal the beak) simultaneously with the blow; **2]** the back with the dorsal fin appears, with the top of the head usually still visible; **3]** the animal rolls languidly and submerges with the tail-stock slightly arched.
Behaviour: does not often breach, and appears awkward when it does, with a vertical leap and an ungainly fall back; often seen logging; an inquisitive species.

Very rare in British waters but regularly seen on summer ferry crossings across the Bay of Biscay. However, there have been many records of strandings, mainly in the west.

Identification: A medium-sized whale, but large for a beaked whale. Rather distinctive, with a reddish-brown body colour; a dorsal fin located two-thirds along the back; and a whitish 'goose-like' head in which the forehead slopes smoothly down to the short but reasonably distinct beak. Adult males often have obvious scarring on the back, recalling Risso's Dolphin (*page 297*) which has a much taller dorsal fin. Juveniles are darker brown and less obviously creamy white about the head.
Habits: Migratory, in British and Irish waters May–September. Usually seen singly, but sometimes in pods of up to 12, which often comprise just females and calves. **Status:** Regular but very rare visitor.

Mesoplodon beaked whales

Four species of beaked whale in the genus *Mesoplodon* have been recorded – three in British and one in Irish waters. These are some of the rarest and most enigmatic of all cetaceans. Squid-feeding denizens of deep, offshore waters, they are unobtrusive and inconspicuous, even when surfacing. Very little is known about their social behaviour, although they have been observed breaching (often multiple times), tail-slapping and porpoising. *Mesoplodon* beaked whales are reminiscent of an elongated dolphin in shape and are small and slim with a small, falcate dorsal fin located two-thirds of the way along the back. The upperparts are typically uniformly brown to grey (males may be scarred), becoming paler towards the belly; the eyes are typically encircled by a dark patch. The rounded forehead tapers to a slender beak, the length of which varies between species. The best feature for identifying *Mesoplodon* species at sea is the position of the teeth (present in adult males only) that protrude from the lower jaw.

SOWERBY'S BEAKED WHALE
ABOVE: *surfacing male showing the beak and the diagnostic position of the teeth.*
BELOW: *a typical view of a Mesoplodon beaked whale – small falcate dorsal fin located two-thirds of the way along the back.*

Blow: low, small, bushy, slightly forwards; inconspicuous. **Dive sequence: 1]** the melon and head appear first, sometimes at a steep enough angle to reveal the beak simultaneously with the blow; **2]** the back ahead of the dorsal fin appears, with the top of the head usually still visible; **3]** the animal rolls smoothly and submerges with the tail-stock slightly arched. **Behaviour:** spends little time at the surface, usually only 1–3 minutes before diving; only breaches occsasionally, and appears awkward when doing so, with a vertical leap and an ungainly fall back into the water; avoids vessels.

Sowerby's Beaked Whale LP | BL (En, Sc)

Mesoplodon bidens

By beaked whale standards, Sowerby's Beaked Whale is almost well-known, with more than 80 records of stranded specimens. Although finding one at sea remains a huge challenge, it is seen relatively frequently in the Faeroe-Shetland Channel, and may possibly be regular in British and Irish waters.

Identification: Typical *Mesoplodon* beaked whale shape. Sowerby's Bealed Whale has a small head, with a rounded forehead and a slender beak, longer than that of the other beaked whales. It is dark grey above, and paler beneath (although some all-dark individuals have been observed) with some pale spots and (in males) scratches on the back and flanks and a dark patch around the eye. A diagnostic feature (on the males only) is the two flattened teeth that protrude midway along the lower jaw.
Status: May breed in British/Irish waters, but poorly known.

Length:	2·1–2·8 m
Weight	135–272 kg

♂

♀

Forehead: Rounded, tapering to long, slender beak; flatter taper in ♀.
Teeth (♂): one, midway along lower jaw.

True's Beaked Whale LP BL (En, Sc)

Mesoplodon mirus

This rare and poorly known cetacean is known from 10 strandings in the west of Ireland and 1 from Sutherland (2020) .

Length:	4·8–5·5 m
Weight	0·9–1·5 t

Identification: A small, lethargic *Mesoplodon*, but obviously larger than a dolphin. Long-bodied, with a stubby, dolphin-like beak and a small, gently sloping bump for a melon. Based on two photographed individuals, body colour can vary from virtually all-dark with a dark-tipped beak, to plain grey above and whitish below (similar to Gervais' Beaked Whale). Both showed a dark patch around the eye with a white corona. Mature males have two small triangular teeth protruding from the very tip of the lower jaw, and a rather straight mouth line. **Status:** Vagrant: 11 records, all strandings.

♂

FOREHEAD: Rounded, tapering to a short, stubby beak.
TEETH (♂): At tip of lower jaw.

Gervais' Beaked Whale LP

Mesoplodon europaeus

A very poorly known deep water species. The first specimen for science was a corpse found floating in the English Channel in 1840. There have been no subsequent records from Great Britain but one was stranded off Co. Sligo, Ireland in January 1989.

Length:	4·0–5·2 m
Weight	1·0–2·6 t

Identification: A rather plump *Mesoplodon*, with a medium-length beak and an indistinct melon. Dark blue-grey above, fading to paler below, often with a dark patch around the eye and a few scars on the back. The only truly diagnostic character is that older males have triangular protruding teeth one-third of the way along the jaw. **Status:** Vagrant: two records of dead animals.

♂

FOREHEAD: Smooth taper to a shortish, slender beak.
TEETH (♂): one-third from tip of lower jaw.

Blainville's Beaked Whale LP

Mesoplodon densirostris

There are just three stranding records of this exceedingly rare beaked whale in Britain: Wales (July 1993), Cornwall (December 2013) and Portland, Dorset (July 2020).

Length:	4·5–6·0 m
Weight	0·7–1·0 t

Identification: Rather plainly coloured, dark brown or dark grey, sometimes with scars on the back. The beak is unique, with a strangely shaped lower jaw arching strongly upwards in a bulge around the base of the beak (more pronounced in males). Males also have teeth that protrude from the arch of the lower jaw. The forehead melon is very small, giving a pointed-head profile. A range of body colours has been observed, from all-dark, all-grey or all-brown, to dark greyish-brown above and pale beneath. Beaks have been all-dark, dark-tipped with brown lower jaws and dark upper jaw with a white lower jaw. **Status:** Vagrant: three records of dead animals.

♂

♀

Forehead: Raised, arched jawline; more pronounced in ♂.
Teeth (♂): erupt from lower jaw.

Pygmy Sperm Whale `LP`

Kogia breviceps

Length:	2·7–3·7 m
Weight	315–450 kg

Blow: low, very light and offset slightly to the left.
Dive sequence: it tends not to dive as such, but instead sinks slowly beneath the surface. It also surfaces and swims slowly.
Behaviour: usually seen logging with the head not visible. Occasionally breaches with a vertical leap.

This curious, deep water toothed cetacean may be numerous far offshore to the west of Britain/Ireland, but is rarely seen.

Identification: Size of a large dolphin, but with a very different shape and habits. It is small but compact with a blunt, squarish head and a very low, hooked dorsal fin set just over halfway along the back behind a distinct hump. It can look quite shark-like, with a dark blue-grey back and a strange pale 'false gill' behind the eye. The creamy underside may have a pink flush. **Habits:** Singles, females with calves or small groups up to six are seen. Lethargic and unobtrusive at the surface, with slow, listless movements; it is not especially shy. If threatened, it may expel a dark reddish-brown liquid from its intestine as a screen to escape. **Status:** Rare, but possibly resident off the Southwest Approaches and Ireland.

Dwarf Sperm Whale `LP`

Kogia sima

Length:	2·1–2·8 m
Weight	135–272 kg

Blow: low, very light. and offset slightly to the left.
Dive sequence: much as Pygmy Sperm Whale, surfaces slowly and sinks slowly, like a submarine.
Behaviour: frequently logs inconspicuously on the surface, motionless; breaching rare – a vertical leap, falling tail or belly first.

This widely distributed but elusive and mysterious animal was recorded for the first time in British waters in Cornwall, in 2011. It is very closely related to the Pygmy Sperm Whale and was only recognised as a species in 1966.

Identification: Extremely similar to Pygmy Sperm Whale, but the key difference is the shape and placement of the dorsal fin – it is prominent, pointed (sometimes with a backward curved tip) and closer to the middle of the back than that of Pygmy Sperm Whale. The back is also flatter in profile and the head is more conical. **Habits:** Found alone or in groups up to ten. **Status:** Vagrant: one record.

Pygmy Sperm Whale has a small, falcate, dorsal fin located more than halfway along the back.

Dwarf Sperm Whale has a relatively broad-based dorsal fin that is located centrally on the back .

False Killer Whale NT `LP` `BL (Sc)`

Pseudorca crassidens

Length:	3·7–6·1 m
Weight	1·01–2·03 t

Blow: bushy, inconspicuous.
Dive sequence: 1] head and blow seen first; **2]** a quick roll shows the back and dorsal fin; **3]** sinks without showing the tail-stock.
Behaviour: lively, sometimes leaping clear of the water in low, flat arcs. Travels at speed in groups of 2–200; known to bow-ride.

Large, sleek, uniformly dark 'blackfish' with prominent dorsal fin, normally found in warm temperate oceans south of 36° N (Gibraltar is the usual northern limit of its range).

Identification: A powerful 'blackfish' dolphin, much larger and more streamlined than other dolphins. The tall, centrally-located dorsal fin is sickle-shaped with a blunt-tip. The slender head is blunt and beakless, with an inconspicuous melon. If seen, the flippers are distinctively bent backwards in the middle. **Habits:** Very sociable, groups ranging from two to 200 (in British waters two pods held 10–20 and 100–150). It has been known to associate with Bottlenose Dolphin (*page 292*). **Status:** Vagrant: mass strandings in the 1920s and 1930s; since 1976 five records (involving multiple animals) as far north as Orkney.

Melon-headed Whale `LP`

Peponocephala electra

Length:	2·0–2·7 m
Weight	up to 210 kg

Blow: low, inconspicuous.
Dive sequence: energetic; porpoising low and seemingly just skimming the surface. Blows as it exits the water.
Behaviour: very low leaps with much splashing, only occasionally breaches.

A fast-moving 'blackfish' dolphin from tropical waters, with one British record – a skull found in Cornwall in 1949.

Identification: Size of a typical dolphin, with a tall, centrally located, sickle-shaped dorsal fin similar to that of Bottlenose Dolphin (*page 292*). However, the entirely dark grey coloration, blunt, rounded head without a protruding beak, and torpedo-shaped body with a thin tail-stock should rule out most dolphins (although beware young Risso's Dolphin, *page 297*). A good view of a Melon-headed Whale might reveal its sharply-pointed flippers, the dark 'cape' on the back around the dorsal fin and black mask around the eye. **Habits:** Very sociable. Subgroups of large pods characteristically split off and change direction. **Status:** Vagrant: one record of a long-dead animal.

Beluga LP

Delphinapterus leucas

Length:	up to 5·5 m (male),
	up to 4·1 m (female)
Weight	400–1,500 kg

Blow: low, inconspicuous.
Dive sequence: 1] low, gentle sur with an inconspicuous blow; **2]** a roll of the white back; **3]** flukes are sometimes lifted and are last to disappear.
Behaviour: tail-slapping and spy-hopping are frequent; breaching rare.

An extraordinary cetacean from the Arctic. One of the few marine mammals which is truly unmistakable, with its remarkable white coloration and lack of any dorsal fin.

Identification: Chalky-white adults are only likely to be mistaken for tiny icebergs, or rare albino individuals of other cetaceans, such as Harbour Porpoise. Medium-sized, Belugas are plump, with a rounded head, no dorsal fin and flukes with a convex trailing edge. Young animals are born mottled grey and get steadily whiter, sometimes with a yellow tinge, with age. **Habits:** Very sociable, pods of 2–10, often single sex. Very playful and lively; often hunts co-operatively. Makes various whistles and trills in communication between members of a pod that can sometimes be heard. **Status:** Vagrant: more than 20 records since 1980 (Shetland, north-east England, north and west Scotland, two off north-east England and one off Co. Antrim, Ireland (2015) and, most recently, one in the River Thames off Gravesend, Kent (2018).

Narwhal LP

Monodon monoceros

Length:	4–5 m
	(plus tusk 2–3 m)
Weight	0·8–1·6 t

Blow: low, inconspicuous.
Dive sequence: smooth; **1]** the head (plus tusk) appearing with the blow; **2]** a roll over with just part of the back breaking the surface (though not the flukes).
Behaviour: lively, but breaching usually incomplete. It will float motionless on the surface in calm conditions, when the tusk may still show.

This remarkable cetacean rarely ventures south of southern Arctic oceans. Tusked males are unmistakable, but there have been no confirmed sighting in Britain since 1949.

Identification: A medium-sized cetacean, with a tusk (in male only) up to 3 m long.. Highly distinctive; stout-bodied with a small, bulbous head and no dorsal fin; the flukes are convex, with a deep notch. Whitish mottled black or grey, with a whiter underside. Mature animals whiten with age; youngsters are more uniform grey. **Habits:** Family (females and calves) groups in summer (2–20); more solitary in winter. Shy of boats. Quite boisterous, and swims fast. **Status:** Vagrant: six old records, from as far south as the Thames Estuary.

Status and legislation

The table on *pages 314–316* provides, for the first time in a popular book, a complete list of the mammal species recorded in Britain and Ireland in modern times, together with a summary of their status and the legislative protection they are currently afforded. The species are listed in taxonomic order, reflecting the apparent natural relationships between families, genera and species, although this is the subject of considerable debate and is constantly changing in the light of new scientific evidence. The order followed is that adopted in *Mammal Species of the World: A Taxonomic and Geographic Reference* (3rd ed.) by D. E. Wilson & D. M. Reeder (editors) (2005), published by Johns Hopkins University Press, except for cetaceans where the treatment adopted by the Society for Marine Mammalogy (2014) (marinemammalscience.org/species-information/ list-marine-mammal-species-subspecies) is followed.

STATUS Key to codes

N **Naturally occurring** species (including rare migrants and vagrants)
I **Introduced** by Man (either deliberately or accidentally)
R **Reintroduced** (native species that became extinct in the region or were extirpated).

SPECIES Key to shading

☐ Established breeders ☐ Feral/semi-domesticated breeders
☐ Ephemeral introductions that have bred.
☐ Rare migrants or vagrants that have occurred naturally

IUCN RED LIST STATUS

The International Union for Conservation of Nature (IUCN) assesses the global conservation status of mammals on a regular basis and maintains a 'Red List'. The categories are summarised below, and are shown for each species in the main table.

EX	Extinct	No reasonable doubt that the last individual has died [category not currently applied to any British/Irish species]
EW	Extinct in the wild	Known only to survive in captivity or as a naturalized population (or populations) well outside the past range.
CR	Critically Endangered	Considered to be facing an extremely high risk of extinction in the wild [category not currently applied to any British/Irish species]
EN	Endangered	Considered to be facing a very high risk of extinction in the wild.
VU	Vulnerable	Considered to be facing a high risk of extinction in the wild.
NT	Near Threatened	Close to qualifying for/likely to qualify as CR, EN or VU in the near future.
LC	Least Concern	Does not qualify as CR, EN, VU or NT.
DD	Data Deficient	Insufficient information for a direct or indirect assessment of the risk of extinction of the species to be made.

(Species categorised as EW, CR, EN and VU are termed 'Globally Threatened'.)

LEGAL PROTECTION

The legal framework that affords protection to mammals in the UK and the Republic of Ireland is complex, principally due to the combination of international, European and domestic legislation, which applies in different ways in the five countries concerned and further changes are likely in future in light of the UK leaving the European Union (EU). Although a detailed overview of the subject is outside the scope of this book, the key current legislation is outlined below. For those species that are afforded protection, a red box is included in the species account and the relevant legislation is summarised in the *List of British and Irish Mammals* that follows.

INTERNATIONAL AND EUROPEAN UNION (EU) LEGISLATION   [INT. LEG.]

The current EU legislative framework provides a logical, evidence-based and robust set of directives that shape national species and habitat protection laws.

Habitats Directive [Hab Dir. Ann.] The principal EU directive that affects mammals is the Habitats Directive (Council Directive 92/43/EEC on the Conservation of Natural Habitats and of Wild Fauna and Flora), which requires Member States to designate protected areas for certain habitats and species, and to protect species listed in various Annexes:

Annex 2 species are those for which their conservation requires the designation of Special Areas of Conservation (SAC);

Annex 4 species are those in need of strict protection (for these, the figure **4** is highlighted in bold text in the *List of British and Irish Mammals* that follows);

Annex 5 species require management measures to be put in place in relation to their being taken in the wild or exploited.

Bern Convention [Bern Con. App.] The Bern Convention (Convention on the Conservation of European Wildlife and Natural Habitats) aims, amongst other things, to ensure the conservation and strict protection of certain animal species and their natural habitats (listed in **Appendix 2**) and to protect and regulate the exploitation of species (including migratory species) listed in **Appendix 3**. It imposes legal obligations on contracting parties, which include the UK and Ireland.

Bonn Convention [Bonn Con.] The Bonn Convention (Conservation of Migratory Species of Wild Animals (CMS)) is a global treaty that aims to conserve terrestrial, marine and avian migratory species throughout their range.

The Environmental Liability Directive (Council Directive 2004/35/EC on environmental liability with regard to the prevention and remedying of environmental damage) is also relevant to some species of mammal. This directive makes those causing damage to the environment, including species protected under the Habitats Directive, legally and financially responsible for that damage.

DOMESTIC LEGISLATION   [DOMESTIC LEG.]

There are many laws that affect mammals in Britain and Ireland, including Acts that relate to specific species or groups of species (*e.g.* Badger, deer, seals). However, the key Act that affords legal protection in England and Wales is the Wildlife and Countryside Act 1981 (as amended) (WCA). In Scotland, most species are protected through Scottish-specific amendments to the WCA (*i.e.* the Nature Conservation (Scotland) Act 2004 and the Wildlife and Natural Environment (Scotland) Act 2011. In Northern Ireland, protection is afforded through the Wildlife (Northern Ireland) Order 1985 (as amended by the Wildlife and Natural

Environment Act (Northern Ireland) 2011. The Conservation of Habitats and Species Regulations 2017 (as amended) continue to apply despite the UK leaving the EU in 2020, due to The Conservation of Habitats and Species Amendment (EU Exit) Regulations 2019. In the Republic of Ireland (É), species are afforded protection under The Wildlife Act 1976 (as amended). The following codes are used in the *List of the mammals of Great Britain and Ireland* on *page 314.*

● Species that are protected at all times.

○ Species that are afforded protection at certain times (*e.g.* deer during a specified closed season) or are protected from being killed or taken by certain methods.

▼ Species that it is an offence to release or allow to escape into the wild. This includes any animal that is not ordinarily resident, or is listed in specific schedules.

In summary, this means:

■ All bats are fully protected and it is an offence to disturb a bat roost intentionally or recklessly, whether or not bats are present at the time. Bats may be handled only under licence (even when dead).

■ All species of cetacean are fully protected.

■ All shrews are protected in the UK and may not be trapped without a licence; when trapping for other small mammals, precaution must be taken to minimise the chances of killing or harming shrews.

■ The other terrestrial mammals that are legally protected in the UK are: Red Squirrel, Hazel Dormouse, Water Vole, Hedgehog, Wildcat, Otter, Pine Marten, Badger and Polecat, and in most places Grey and Common Seal.

■ The following species are protected in Ireland: Red Squirrel, Mountain/Irish Hare, Hedgehog, Pygmy Shrew, Badger, Otter, Pine Marten, Stoat and all species of bat, deer, seal and cetacean.

■ It is now illegal to hunt any animal with dogs in England, Scotland and Wales. This relates to Fox, Badger, Otter and Brown Hare, among others.

The process of obtaining a licence depends on the country concerned. For information, contact the relevant Statutory Nature Conservation Body (SNCB) (see *page 318*).

This section has been written as a non-technical summary.
Anyone requiring specific legal advice regarding the protection of mammals should consult a qualified legal professional.

BIODIVERSITY LIST [BIODIVERSITY LIST]

Under the UK Biodiversity Action Plan (UK BAP), species and habitats are highlighted for conservation action, and lists have been drawn up for each country. Lists of these habitats and species are published as a requirement of Section 41 (England) and Section 42 (Wales) of the Natural Environment and Rural Communities (NERC) Act 2006, Section 2(4) of the Nature Conservation (Scotland) Act 2004, and Section 3(1) of the Wildlife and Natural Environment Act (Northern Ireland) 2011. They are used to guide decision-makers, such as local and regional authorities, in their duty "to have regard to the conservation of biodiversity in the exercise of their normal functions". ● indicates that the species is included on the relevant country's Biodiversity List: England (En); Wales (Wa); Scotland (Sc) and Northern Ireland (NI). In the Republic of Ireland (É), Biodiversity Action Plans are largely based on counties or local authority areas.

List of the mammals of Great Britain and Ireland

STATUS	SPECIES	IUCN RED LIST	INT. LEG. Hab Dir. Ann.	INT. LEG. Bern Con. App.	INT. LEG. Bonn Con.	DOMESTIC LEG. UK	DOMESTIC LEG. É	BIODIVERSITY LIST NERC En	BIODIVERSITY LIST NERC Wa	BIODIVERSITY LIST Biod. List Sc	BIODIVERSITY LIST Biod. List NI	Page
I	Red-necked Wallaby	LC				▼						261
I	Black-tailed Prairie-dog	LC				▼						263
I	Grey Squirrel	LC				▼						58
N	Red Squirrel	LC		3		●	●	●	●	●	●	54
I	Siberian Chipmunk	LC				▼						263
I	Edible (Fat) Dormouse	LC		3		▼						62
N	Hazel Dormouse	LC	4	3		●		●	●			64
R	Eurasian Beaver	LC	2, 4	3		▼						98
N	Water Vole	LC				●		●	●	●		76
I	Golden Hamster	VU				▼						265
N	Field Vole	LC										72
I	Orkney (Common) Vole	LC								[●]		74
N	Bank Vole	LC										69
I	Muskrat	LC				▼						265
N	Yellow-necked Mouse	LC										88
N	Wood Mouse	LC										84
I	Mongolian Gerbil	LC				▼						264
N	Harvest Mouse	LC						●	●			80
I	House Mouse	LC										90
I	Common Rat	LC				▼†						92
I	Ship (Black) Rat	LC				▼				●		96
I	Himalayan Porcupine	LC				▼						266
I	Coypu	LC				▼						264
N	Brown Hare	LC				▼‡	●	●	●	●		120
N	Mountain/Irish Hare	LC	5	3		●*	●	●		●	●	124
I	Rabbit	LC				▼						116
N	Hedgehog	LC		3		●▼†	●	●	●	[●]	●	112
I	Greater White-toothed Shrew	LC		3		●						108
I	Lesser White-toothed Shrew	LC		2, 3		●						106
N	Water Shrew	LC		3		●						104
N	Common Shrew	LC		3		●						100
N	Pygmy Shrew	LC		3		●	●					102
N	Mole	LC										110
N	Greater Horseshoe Bat	LC	2, 4	2	●	●		●	●	●		140
N	Lesser Horseshoe Bat	LC	2, 4	2	●	●		●	●	●		144
N	European Free-tailed Bat	LC	4	2	●	●						190
N	Barbastelle	NT	2, 4	2	●	●		●	●			162
N	Northern Bat	LC	4	2	●	●						186
N	Serotine	LC	4	2	●	●						146

* Scotland only
‡ Northern Ireland only
† With respect to introductions onto islands in territorial waters in Northern Ireland only
square brackets [●] signifies 'watching brief' only

| STATUS | SPECIES | IUCN RED LIST | INT. LEG. | | | DOMESTIC LEG. | | BIODIVERSITY LIST | | | | Page |
| | | | Hab Dir. Ann. | Bern Con. App. | Bonn Con. | UK | É | NERC | | Biod. List | | |
								En	Wa	Sc	NI	
N	Savi's Pipistrelle	LC	4	2	●	●						189
N	Hoary Bat	LC		2		●						191
N	Alcathoe Bat	DD	4	2	●	●						180
N	Bechstein's Bat	NT	2, 4	2	●	●		●	●			170
N	Brandt's Bat	LC	4	2	●	●	●			●		182
N	Pond Bat	NT	2, 4	2	●	●						193
N	Daubenton's Bat	LC	4	2	●	●	●			●		174
N	Geoffroy's Bat	LC	2, 4	2	●	●						192
N	Greater Mouse-eared Bat	LC	2, 4	2	●	●						184
N	Whiskered Bat	LC	4	2	●	●	●			●		178
N	Natterer's Bat	LC	4	2	●	●	●			●		172
N	Leisler's Bat	LC	4	2	●	●	●					152
N	Noctule	LC	4	2	●	●		●	●	●		148
N	Kuhl's Pipistrelle	LC	4	2	●	●						188
N	Nathusius' Pipistrelle	LC	4	2	●	●	●			●	●	160
N	Common Pipistrelle	LC	4	3	●	●			●	●		154
N	Soprano Pipistrelle	LC	4	2	●	●	●	●	●	●	●	158
N	Brown Long-eared Bat	LC	4	2	●	●	●	●	●	●	●	164
N	Grey Long-eared Bat	LC	4	2	●	●						168
N	Particoloured Bat	LC	4	2	●	●						187
I	Feral Cat	LC										201
N	Wildcat	LC	4	2		●				●		198
N	Fox	LC				▼†						202
N	Walrus	VU	5	2		●						275
N	Hooded Seal	VU	5	3								276
N	Bearded Seal	LC		3								277
N	Grey Seal	LC	2, 5	3	●	●○	●					272
N	Harp Seal	LC		3								278
N	Ringed Seal	LC		3								279
N	Common Seal	LC	2, 5	3	●	●○	●	●		●	●	270
I	Asian Short-clawed Otter	LC				▼						267
N	Otter	NT	2, 4	2		●	●	●	●		●	210
N	Pine Marten	LC	5	3		●	●	●	●		●	216
N	Badger	LC		3		●‡	●					206
N	Stoat	LC		3		▼†	●					220
N	Weasel	LC		3								224
N	Polecat	LC	5	3		●		●	●			218
I	American Mink	LC				▼						214
I	Striped Skunk	LC				▼						266
I	Raccoon	LC				▼						267
I	Horse	LC										258

‡ Also protected in the UK under The Protection of Badgers Act 1992
† With respect to introductions onto islands in territorial waters in Northern Ireland only

STATUS	SPECIES	IUCN RED LIST	INT. LEG. Hab Dir. Ann.	Bern Con. App.	Bonn Con.	DOMESTIC LEG. UK	É	BIODIVERSITY LIST NERC En	Wa	Biod. List Sc	NI	Page
R	Wild Boar	LC				▼						226
N	European Roe Deer	LC		3		○▼‡						230
I	Sika	LC		3		○▼	●○					238
N	Red Deer	LC		3		○	●○					234
I	Fallow Deer	LC		3		○	●○					242
I	Père David's Deer	EW		3		▼						262
I	Chinese Water Deer	VU		3		○▼						248
I	Reeves' Muntjac	LC		3		▼						246
R	Reindeer	VU		3								251
I	Cattle	LC										256
I	Feral Goat	LC										252
I	Feral Sheep	LC										254
N	Bowhead Whale	LC	4	3	●	●	●					303
N	North Atlantic Right Whale	CR	4	3	●	●	●	●		●		303
N	Minke Whale	LC	4	2	●	●	●	●	●	●	●	282
N	Sei Whale	EN	4	2	●	●	●	●		●	●	300
N	Blue Whale	EN	4	2	●	●	●			●		304
N	Fin Whale	VU	4	2	●	●	●	●	●	●		284
N	Humpback Whale	LC	4	2	●	●	●		●	●	●	286
N	Sperm Whale	VU	4	2	●	●	●	●		●		301
N	Pygmy Sperm Whale	LC	4	2	●	●	●					308
N	Dwarf Sperm Whale	LC	4	2	●	●	●					308
N	Northern Bottlenose Whale	DD	4	2	●	●	●		●	●		302
N	Sowerby's Beaked Whale	LC	4	2	●	●	●	●		●		306
N	Gervais' Beaked Whale	LC	4	3	●	●	●			●		307
N	True's Beaked Whale	LC	4	2	●	●	●	●		●		307
N	Blainville's Beaked Whale	LC	4	2	●	●	●			●		307
N	Cuvier's Beaked Whale	LC	4	2	●	●	●	●	●	●		305
N	Beluga	LC	4	3	●	●	●					310
N	Narwhal	LC	4	2	●	●	●					310
N	Short-beaked Common Dolphin	LC	4	2	●	●	●	●	●	●	●	294
N	Long-finned Pilot Whale	LC	4	2	●	●	●	●	●	●		290
N	Risso's Dolphin	LC	4	2	●	●	●	●	●	●		297
N	Fraser's Dolphin	LC	4	3	●	●	●			●		305
N	Atlantic White-sided Dolphin	LC	4	2	●	●	●	●	●	●		295
N	White-beaked Dolphin	LC	4	2	●	●	●	●	●	●		296
N	Killer Whale (or Orca)	DD	4	2	●	●	●	●	●	●	●	288
N	Melon-headed Whale	LC	4	3	●	●	●			●		309
N	False Killer Whale	NT	4	2	●	●	●			●		309
N	Striped Dolphin	LC	4	2	●	○▼	●		●	●		299
N	Bottlenose Dolphin	LC	2, 4	2	●	●	●	●	●	●	●	292
N	Harbour Porpoise	LC	2, 4	2	●	●	●	●	●	●	●	298

‡ Northern Ireland only

Further reading and sources of useful information

Recommended reference books

Atkinson, R. 2013. *Moles*. Whittet Books.

Aulangier, S. *et al.* 2009. *Mammals of Europe, North Africa and the Middle East.* Christopher Helm.

Birks, J. 2015. *Polecats*. Whittet Books.

Briggs, B. & King, D. 1998. *The Bat Detective*. Batbox Ltd.

Carter, P. & Churchfield, S. 2006. *The Water Shrew Handbook.* The Mammal Society.

Chanin, P. 2013. *Otters*. Whittet Books.

Clark, M. 1994 (2010). *Badgers*. Whittet Books.

Dietz, C. *et al.* 2016. *Bats of Britain and Europe.* Bloomsbury Publishing.

Dunn, J., Still, R. & Harrop, H. 2012. *Britain's Sea Mammals.* Princeton WILD*Guides*.

Gurnell, J. & Flowerdew, J.R. 2006. *Live Trapping Small Mammals: A Practical Guide.* The Mammal Society.

Gurnell, J., Lurz, P.W.W. & Wauters, L. 2012. *Squirrels*. The Mammal Society.

Harris, S. & Yalden, D.W. (eds.) 2008. *Mammals of the British Isles: Handbook, 4th edition*. The Mammal Society.

Macdonald, D. & Barrett, P. 2005. *Collins Field Guide to Mammals of Britain and Europe.* HarperCollins.

Macdonald, R. & Harris, S. 2006. *Stoats and Weasels*. The Mammal Society.

Moores, R. 2007. *Where to Watch Mammals: Britain and Ireland.* A&C Black.

Morris, P.A. 2011. *Dormice: A Tale of Two Species*. Whittet Books.

Morris, P.A. 2011. *The Hedgehog*. The Mammal Society.

Morris, P.A. 2014. *Hedgehogs*. Whittet Books.

Muir, G. & Morris, P.A. 2013. *How to Find and Identify Mammals.* The Mammal Society.

Olsen, L-H. 2013. *Tracks and Signs of the Animals and Birds of Britain and Europe.* Princeton University Press.

Richardson, P. 2000. *Bats*. Whittet Books.

Russ, J. 2012. *British Bat Calls: A Guide to Species Identification*. Pelagic Publishing.

Sterry, P. 2010. *Collins Complete Guide to British Animals*. HarperCollins Publishers.

Still, R., Harrop, H., Stenton, T. & Diaz, L. 2019. *Europe's Sea Mammals*. Princeton WILDGuides.

Strachan, R. 2010. *Mammal Detective*. Whittet Books.

Tapper, S & Yalden, D. 2010. *The Brown Hare*. The Mammal Society.

Waters, W & Warren, R. 2009. *Bats*. The Mammal Society.

Wembridge, D. & Bowen, C.P. 2010. *Britain's Mammals: A Concise Guide.* Whittet Books.

Wilson, D.E. & Mittermeier, R.A. (eds.) 2009-16. *Handbook of the Mammals of the World, Volumes 1–6.* Lynx Edicions.

Woodroffe, G. 2007. *The Otter*. The Mammal Society.

Woods, M. 2010. *The Badger*. The Mammal Society.

Yalden, D.W. 1999. *The History of British Mammals*. T&AD Poyser (A&C Black).

Yalden, D.W. 2009. *The Analysis of Owl Pellets.* The Mammal Society.

Journals

British Wildlife
Independent bi-monthly magazine. Website: britishwildlife.com

Mammal Communications
Online: short papers focused on British and European mammals.
Website: mammal.org.uk/science-research/mammal-communications

Mammal Review
Quarterly journal of the Mammal Society, published by Wiley.
Website: mammal.org.uk/science-research/mammal-review

Online resources

Atlas of Irish Mammals
Website: mammals.biodiversityireland.ie

Bat Conservation Trust
Website: bats.org.uk.

BatCRU (Bat Conservation and Research Unit)
Facebook page: facebook.com/batcru/?ref=aymt_homepage_panel

Birdguides
Website: birdguides.com

Dormice – National Dormouse Monitoring Programme.
Website: surveydata.ptes.org/dormousemonitoring

The Mammal Society
Website: mammal.org.uk

Wild Boar
Website: britishwildboar.org.uk

Training courses

Field Studies Council
Website: field-studies-council.org

The Mammal Society
Website: mammal.org.uk

Statutory Nature Conservation Bodies (SNCBs)

ENGLAND
Enquiries team, Natural England, County Hall, Spetchley Road, Worcester WR5 2NP. Email: enquiries@naturalengland.org.uk. Tel. 0300 060 3900.

WALES
Natural Resources Wales, c/o Customer Care Centre, Ty Cambria, 29 Newport Rd, Cardiff CF24 0TP.
Email: enquiries@naturalresourceswales.gov.uk. Tel. 0300 065 3000.

SCOTLAND
NatureScot Headquarters, Great Glen House, Leachkin Road, Inverness IV3 8NW. Email: ENQUIRIES@Nature.scot. Tel. 01463 725000.
For enquiries regarding seals and some areas of whale and dolphin licensing and fisheries, contact
Marine Scotland, Mailpoint 11, 1B South Victoria Quay, Edinburgh EH6 6QQ. Email: marinescotland@gov.scot. Tel. 0300 244 4000.

NORTHERN IRELAND
Department of Agriculture, Environment and Rural Affairs (DAERA), Dundonald House, Upper Newtownards Road, Ballymiscaw, Belfast BT4 3SB. Email: daera.helpline@daera-ni.gov.uk. Tel. 0300 200 7852.

REPUBLIC OF IRELAND
National Parks and Wildlife Service (NPWS), 7 Ely Place, Dublin 2, Ireland D02 TW98. Email: nature.conservation@ahg.gov.ie. Tel. 1890 383 000 (within Republic of Ireland); +353-1-888 3242 (from UK). Website: npws.ie.

Non-native species

There are excellent sources of information available on the web concerning non-native species, perhaps the most useful and up-to-date websites being those run by the **GB Non-native Species Secretariat** nonnativespecies.org and **Invasive Species Ireland** invasivespeciesireland.com. The website **'Alien Invaders'** hows.org.uk/inter/birds/exotics.htm also provides useful information. Any sightings of non-native 'exotic' species in the wild should be reported: in Great Britain online at nonnativespecies.org/index.cfm?sectionid=81 or by email to the **Centre for Ecology & Hydrology** alertnonnative@ceh.ac.uk; in Ireland to **Invasive Species Ireland** via invasivespeciesireland.com/alien-watch.

Non-statutory conservation organisations

THE MAMMAL SOCIETY   Website: mammal.org.uk

This book has been produced in association with the Mammal Society, a charity that promotes science-led mammal conservation and leads efforts to collect and share information on mammals, encourage research to learn more about their ecology and distribution, and contribute to efforts to conserve them. The Society's free Mammal Tracker app, and online recording forms provide an easy way for the public to submit information on their sightings of mammals. The Society supports a network of experts and enthusiasts who undertake work to survey, monitor, research and conserve mammals, and campaigns for conservation that benefits mammals. It aims to:

- raise public awareness of mammals, their ecology and conservation needs;
- encourage people to participate effectively in mammal monitoring and recording;
- develop the knowledge and field skills of those working for mammals;
- encourage and disseminate the results of research and new information; and
- provide up-to-date, reliable information and science-led advice.

Distribution maps The Mammal Society has recently completed the first review of British terrestrial mammals for 20 years, funded by the statutory nature conservation bodies (SNCB) Natural England, Natural Resources Wales and Scottish Natural Heritage (now NatureScot). The results are published in the *Atlas of Mammals in Great Britain and Northern Ireland**, which are summarized in *Britain's Mammals 2018: The Mammal Society's Guide to their Population and Conservation Status***. The map data in these publications were used as the basis of the distribution maps for Great Britain and Northern Ireland in this book. (See *page 53* for further details.)

* **Crawley, D., Coomber, F., Kubasiewicz, L.M., Harrower, C.A., Evans, P., Waggitt, J., Smith, B., Matthews, F.** (eds.) 2020. *Atlas of the Mammals of Great Britain and Northern Ireland*. Pelagic Publishing.

** **Mathews, F., Coomber, F., Wright, J. & Kendall, T.** (eds.) 2018. *Britain's Mammals 2018: The Mammal Society's Guide to their Population and Conservation Status*. The Mammal Society.

Bat Conservation Trust
Website: bats.org.uk
For information on **local bat groups**, the **National Bat Monitoring Programme** and '**Bat Helpline**': Tel. +44 (0)345 1300 228.

Irish Whale and Dolphin Group
Website: iwdg.ie

Marine Conservation Society
Website: mcsuk.org

ORCA (Organisation Cetacea)
Website: orcaweb.org.uk

People's Trust for Endangered Species (PTES)
Website: ptes.org

Sea Watch Foundation
Website: seawatchfoundation.org.uk.

The Wildlife Trusts
Website: wildlifetrusts.org

Vincent Wildlife Trust
Website: vwt.org.uk

Whale and Dolphin Conservation
Website: uk.whales.org

Acknowledgements and photographic/artwork credits

Many people have contributed, directly or indirectly, to the production of this book, and our sincere thanks go to everyone who has influenced the final product. We would, however, especially like to thank the following: Jen Brumfield for preparing the illustrations of mammal tracks on *pages 46–52*; Adrian Bayley, for his contribution to the bat identification key and comments on the bat and small mammal species accounts; Fiona Mathews and Laura Kubasiewicz at the Mammal Society for their help in facilitating the collation of the maps and the very constructive comments on the text from Derek Crawley and Paul Chanin; Tim Hounsome at the ecological consultancy RSK Biocensus (biocensus.co.uk) for his assistance with the status and legislation section; Kate Jeffreys at the ecological consultancy Geckoella (geckoella.co.uk) for her invaluable input on bat sounds, and for kindly preparing the bat sonograms that appear on *pages 134–135*; John Morgan for generously contributing the information and illustrations of bat droppings that appear on *page 139*; and Daniel Whitby of the ecological consultancy AEWC for his insightful comments on some of the finer points of bat identification. Thanks are also due to Alan Buckle, Tony Cowton, Mike and Penny Hounsome, Rob Hume, Duncan Jones, Tony Mitchell-Jones, Sam Olney, Simon Poulton, Colin Prescott and Dave and Sue Smallshire for their helpful contributions in various ways.

A particular mention must also go to Rachel Still and Gill Swash for their invaluable help in bringing this book to fruition, to Chris and Judith Gibson for providing technical advice and for their editorial input, and to Brian Clews for his help with sourcing images and his keen eye in proof-reading. We also thank Nick Baker for kindly writing the Foreword to the book, and Robert Kirk, Publisher, Field Guides & Natural History at Princeton University Press, for his support and encouragement throughout this project.

The production of this book would not have been possible without the generous support of the many photographers who kindly supplied their images. In total, 500 images are featured, representing the work of 139 photographers. Collating these images proved to be a very time-consuming task and we would like to express our gratitude to Marc Guyt and Roy de Haas at the Agami photo agency in the Netherlands (agami.nl) for their invaluable help in this process.

A number of photographers generously provided access to their entire portfolio of images. Their work is featured extensively throughout the book and their skill is clear to see. All of them spend many hours with a camera, using their technical expertise, extensive local knowledge and understanding of their subject in pursuit of the perfect picture. In this respect we are particularly grateful to Greg and Yvonne Dean (WorldWildlifeImages.com), Andy Harmer (flickr.com/photos/38477281@N05), Hugh Harrop (shetlandwildlife.co.uk), Mark Hows (flickr.com/photos/markhows), Paul Kennedy, David Kjaer (davidkjaer.com), Mike Read (mikeread.co.uk), Roger Tidman, Colin Varndell (colinvarndell.co.uk), Daniel Whitby and Phil Winter (flickr.com/photos/philwinter).

The contribution of every photographer is gratefully acknowledged and each image is listed in this section, together with the photographer's name and website, where appropriate. A number of images have been sourced through the generous terms of the Creative Commons Attribution-ShareAlike 2.0 Generic license. These are indicated by "/CC" after the photographer's name in the list.

Cover: Red Squirrel [David Kjaer (davidkjaer.com)].

Title page: Brown Hare [Roger Tidman].

INTRODUCTION **p8: Greater White-toothed Shrew** [Theo Douma (Agami.nl)]. **p11: Red Deer** [Nigel Cattlin (FLPA)]; **Harvest Mouse** [J-L Klein and M-L Hubert (FLPA)]. **p13: Brown Bear** [Volodymyr Burdiak (Shutterstock)]; **Eurasian Lynx** [Rudmer Zwerver (Shutterstock)]. **p15: Wolf** [Eric Kilby/CC].
p18: Raccoon [Alex O'Neal/CC]; **Red-necked Wallaby** [Sean Kelleher/CC]; **Coypu** [Steven Byrnes/CC].
p19: Red Squirrel [Peter G Trimming/CC]; **Hazel Dormouse** [Danny Green (Agami.nl)]; **Bank Vole** [Theo Douma (Agami.nl)]; **Wood Mouse** [Phil Winter (flickr.com/photos/philwinter)]; **Eurasian Beaver** [Per Harald Olsen (Wikipedia: commons.wikimedia.org/wiki/Castor_fiber#/media/File:Beaver_pho34.jpg)].
p20: Brown Hare [Sergey Yeliseev/CC]; **Hedgehog** [David Kjaer (davidkjaer.com)]; **Common Shrew** [Phil Winter (flickr.com/photos/philwinter)]; **Mole** [Josef Hlasek (Hlasek.com)]. **p21: Greater Horseshoe Bat** [Theo Douma (Agami.nl)]; **Brown Long-eared Bat** [Andy Harmer (flickr.com/photos/38477281@N05)]; **European Free-tailed Bat** [Leonardo Ancillotto (flickr.com/photos/leonardoancillotto)].

p22: Wildcat hybrid [James Lowen (pbase.com/james_lowen)]; **Fox** [Martin Bennett]; **Polecat** [Greg and Yvonne Dean (WorldWildlifeImages.com)]; **Common Seal** [Roger Tidman]; **Walrus** [Polar Cruises/CC]. **p23: Exmoor pony** [Roger Marks (flickr.com/photos/rpmarks)]; **Wild Boar** [David Kjaer (davidkjaer.com)]; **Red Deer** [Andy Morffew/CC]; **Feral Goat** [Mike Read (mikeread.co.uk)]. **p24: Bowhead Whale** [Bering Land Bridge National Preserve/CC]; **Minke Whale** [Hugh Harrop (shetlandwildlife.co.uk)]; **Sperm Whale** [Hugh Harrop (shetlandwildlife.co.uk)]; **Dwarf Sperm Whale** [Glenn Overington]. **p25: Cuvier's Beaked Whale** [Hugh Harrop (shetlandwildlife.co.uk)]; **Beluga** [Hugh Harrop (shetlandwildlife.co.uk)]; **Short-beaked Common Dolphin** [Hugh Harrop (shetlandwildlife.co.uk)]; **Harbour Porpoise** [Hilary Chambers (flickr.com/people/touch_of_frost)]. **p26: Grey Squirrel** [Roger Tidman (FLPA)]; **Red Deer** [Han Bouwmeester (Agami.nl)]. **p27: Water Vole** [Andy and Gill Swash (WorldWildlifeImages.com)]; **Rabbit** [Andy and Gill Swash (WorldWildlifeImages.com)]. **p28: Badger** [Smabs Sputzer/CC]. **p29: Red Deer** [Danny Green (Agami.nl)]. **p30: Mountain Hare** [Hugh Harrop (shetlandwildlife.co.uk)]. **p31: Brown Hares** [Roger Tidman]. **p32: Hedgehog** [Piotr Krzeslak (Shutterstock)]; **Killer Whales** [Hugh Harrop (shetlandwildlife.co.uk)]. **p33: Foxes** [Phil Winter (flickr.com/photos/philwinter)]. **p34: Grey Seal** [James Lowen (pbase.com/james_lowen)]. **p35: Small mammal traps** [Sue Smallshire]. **p36: Common Pipistrelle** [Theo Douma (Agami.nl)]. **p37:** all [Dave Smallshire]. **p38: Whale-watching from land** [Hugh Harrop (shetlandwildlife.co.uk)]. **p39: Whale-watching from a boat** [Natural Resources Wales]. **p40: Rabbit tracks** [Colin Varndell (colinvarndell.co.uk)]; **Otter tracks** [Roger Tidman]. **p41: Vole nest** [Mark Hows (flickr.com/photos/markhows)]; **Badger sett** [Dominic Couzens]. **p42: Droppings: Rabbit** [Andy and Gill Swash (WorldWildlifeImages.com)]; **European Roe Deer** [Andy and Gill Swash (WorldWildlifeImages.com)]; **Common Rat** [Vertebrate Pests Unit, The University of Reading]; **Fox** [Andy and Gill Swash (WorldWildlifeImages.com)]; **Hedgehog** [Roger Tidman]. **p43: Droppings: Otter** [Roger Tidman]; **Badger** [Trevor Codlin]; **Stoat** [Roger Tidman]; **deer**, all [Roger Tidman]. **p44: Gnawed hazelnuts**, all [Roger Tidman]; **Grey Squirrel diggings** [Roger Tidman]; **Badger diggings** [Andy and Gill Swash (WorldWildlifeImages.com)]. **p45: Badger hair** [Andy and Gill Swash (WorldWildlifeImages.com)]; **owl pellet** [Francis Hickenbottom]; **Red Deer** [Mike Read (mikeread.co.uk)]. **pp46–52: Tracks**, all [Jen Brumfield].

RODENTS AND SHREWS p55: Red Squirrel: summer [Andy and Gill Swash (WorldWildlifeImages.com)]; winter [Mike Read (mikeread.co.uk)]. **p56: Red Squirrel** [Greg and Yvonne Dean (WorldWildlifeImages.com)]. **p57: Red Squirrel** [Jules Cox (FLPA)]; drey [Laurie Campbell (lauriecampbell.com)]. **p59: Grey Squirrel**: summer [likeaduck/CC]; winter [Roger Tidman]. **p60: Grey Squirrel** (black) [Ron Rowan Photography (Shutterstock)]; **hazelnut** [Roger Tidman]; **pine cone** [Dominic Couzens]. **p61: Grey Squirrel** [Greg and Yvonne Dean (WorldWildlifeImages.com)]; drey [Pete Adey]. **p62: Edible Dormouse** [Derek Middleton (FLPA)]. **p63: Edible Dormouse** [David Kjaer (davidkjaer.com)]. **p64: Hazel Dormouse** [David Kjaer (davidkjaer.com)]. **p65: Hazel Dormouse**: climbing [Danny Green (Agami.nl)]; asleep [Roger Tidman]; hazelnut [Roger Tidman]. **p66: Hazel Dormouse** [Danny Green (Agami.nl)]; **Common Rat** [David Kjaer (davidkjaer.com)]; **Bank Vole** [Lubomir Hlasek (Hlasek.com)]; **Wood Mouse** [Phil Winter (flickr.com/photos/philwinter)]; **Common Shrew** [Erni (Shutterstock)]. **p67: Common Rat** [Mike Read (mikeread.co.uk)]; **Water Vole** [Andy and Gill Swash (WorldWildlifeImages.com)]. **p68: Common Shrew:** jaw [Roger Tidman]; animal [Colin Varndell (colinvarndell.co.uk)]; **Pygmy Shrew** [Theo Douma (Agami.nl)]; **Water Shrew** [Mike Read (mikeread.co.uk)]; **Greater White-toothed Shrew** [Theo Douma (Agami.nl)]; dentition illustrations [Robert Still]. **p69: Bank Vole** [Colin Varndell (colinvarndell.co.uk)]. **p70:** Hazelnut [Roger Tidman]; **Skomer Vole** [Mike Lane (FLPA)]. **p71: Bank Vole**: dark [Phil Winter (flickr.com/photos/philwinter)]; greyish [Lubomir Hlasek (Hlasek.com)]; chestnut [Roger Tidman]. **p72: Field Vole** [Derek Middleton (FLPA)]. **p73: Field Vole** runs [Jim Barton [CC BY-SA 2.0 (http://creativecommons.org/licenses/by-sa/2.0)], via Wikimedia Commons]. **p74: Orkney Vole** [Martin B Withers (FLPA)]. **p75: Orkney Vole** [pauljennywilson/CC]. **p76: Water Vole** [Erni (Shutterstock)]. **p77: Water Vole** [Greg and Yvonne Dean (WorldWildlifeImages.com)]. **p78: Water Vole** [Andy and Gill Swash (WorldWildlifeImages.com)]. **p79: Water Vole**, both [Andy and Gill Swash (WorldWildlifeImages.com)]. **p80: Harvest Mouse** [Mark Bridger (Shutterstock)]. **p81: Harvest Mice** [Paul Tymon (Shutterstock)]. **p82: Harvest Mouse**: pale [Roger Tidman]; dark [Phil Winter (flickr.com/photos/philwinter)]. **p83: Harvest Mouse**: in nest [Roger Tidman]; leaving nest [Theo Douma (Agami.nl)]. **p84: Wood Mouse** [Phil Winter (flickr.com/photos/philwinter)]. **p85: Wood Mouse** [Theo Douma (Agami.nl)]; hazelnuts [Roger Tidman]. **p86: Yellow-necked Mouse** [David Kjaer (davidkjaer.com)]; **Wood Mouse** [Roger Tidman]. **p87: Yellow-necked and Wood Mice**, both [Mark Hows (flickr.com/photos/markhows)]. **p88: Yellow-necked Mouse** [Phil Winter (flickr.com/photos/philwinter)]. **p89: Yellow-necked Mouse** [David Kjaer (davidkjaer.com)]. **p90: House Mouse** [Theo Douma (Agami.nl)]. **p91: House Mouse** droppings [NY State IPM Program at Cornell University/CC]. **p92: Common Rat** [David Kjaer (davidkjaer.com)]. **p93: Common Rat** [Mike Read (mikeread.co.uk)]. **p94: Common Rats** [Roger Tidman]. **p95: Common Rat**: in hole [David Kjaer (davidkjaer.com)]; swimming [Mike Read (mikeread.co.uk)]. **p96: Black Rat** [Milos Anděra (naturephoto-cz.com)]. **p97: Common Rat** [Erni (Shutterstock)]; **Black Rat** [Milos Anděra (naturephoto-cz.com)].

p98: Eurasian Beaver [Jacques van der Neut (Agami.nl)]. **p99: Eurasian Beaver** swimming [Harvey van Diek (Agami.nl)]; gnawed tree trunk [Alexandre Dulaunoy/CC]. **p100: Common Shrew**: jaw [Roger Tidman]; feeding [Theo Douma (Agami.nl)]. **p101: Common Shrew** [Erni (Shutterstock)]; dentition illustration [Robert Still]. **p102: Pygmy Shrew** [Theo Douma (Agami.nl)]. **p103: Pygmy Shrew** [Phil Winter (flickr.com/photos/philwinter)]; dentition illustration [Robert Still]. **p104: Water Shrew** [Lubomir Hlasek (Hlasek.com)]. **p105: Water Shrew** [Erni (Shutterstock)]; dentition illustration [Robert Still]. **p106: Lesser White-toothed Shrew** [Josef Hlasek (Hlasek.com)]. **p107: Shrew** dentition illustrations [Robert Still]. **pp108 & 109: Greater White-toothed Shrew**, both [Theo Douma (Agami.nl)].

MOLE, HEDGEHOG, RABBIT AND HARES p110: Mole [Josef Hlasek (Hlasek.com)]. **p111: Mole** [Lubomir Hlasek (Hlasek.com)]; molehills [Andy and Gill Swash (WorldWildlifeImages.com)]. **p112: Hedgehog** [Roger Tidman]. **p113: Hedgehog**: rolled-up [Laurie Campbell (lauriecampbell.com)]; walking [Wil Leurs (Agami.nl)]. **p114: Hedgehog** [Roger Tidman]. **p115: Hedgehog** [Miroslav Hlavko (Shutterstock)]; hibernaculum [Andy and Gill Swash (WorldWildlifeImages.com)]. **p116: Rabbit** [Greg and Yvonne Dean (WorldWildlifeImages.com)]. **p117: Rabbit**: black [Roger Tidman]; crouched [James Lowen (pbase.com/james_lowen)]. **p118: Rabbit** [Roger Tidman]. **p119: Rabbit** [Erni (Shutterstock)]; warren [Nigel Kenyon Jones]. **p120: Brown Hare** [Mike Read (mikeread.co.uk)]. **p121: Brown Hare** [Mike Read (mikeread.co.uk)]. **p122: Brown Hare** [Mike Read (mikeread.co.uk)]. **p123: Brown Hare**: boxing [bikeriderlondon (Shutterstock)]; in form [Erica Olsen (FLPA)]. **p125: Mountain Hare**: summer [Ralph Todd]; winter [Roger Tidman]. **p126: Irish Hare** [[John M Kelly (flickr.com/photos/rosefactor). **p127: Mountain Hare**: summer [Hugh Harrop (shetlandwildlife.co.uk)]; moulting [David Kjaer (davidkjaer.com)].

BATS p129: Greater Horseshoe Bat [Dave Smallshire]; **Soprano Pipistrelle** [Laurie Campbell (lauriecampbell.com)]; **Noctule and Nathusius' Pipistrelles** [Theo Douma (Agami.nl)]; **Brandt's Bat** [Paul Kennedy]. **p130:** Nose-leaf [Andy and Gill Swash (WorldWildlifeImages.com)]; others [Theo Douma (Agami.nl)]. **p131: Pipistrelle** wing illustrations [Robert Still]; post-calcarial lobe absent [Andy Harmer (flickr.com/photos/38477281@N05)]; long-eared bat thumbs, both [Daniel Whitby]; others [Theo Douma (Agami.nl)]. **p132:** Calcar: short [Andy Harmer (flickr.com/photos/38477281@N05)]; long [Theo Douma (Agami.nl)]; *Myotis* illustrations, all [Robert Still]. **pp134–135: Bat sonograms**, all [Kate Jeffreys (Geckoella.co.uk)]. **p138: Droppings: Common Pipistrelle** [Hugh Clark (FLPA)]; **Brown Long-eared Bat** [Trevor Codlin; **Noctule roost in tree** [Theo Douma (Agami.nl)]. **p139: Bat droppings**, all illustrations [John Morgan]. **pp141–142: Greater Horseshoe Bat**: face from front [Trevor Codlin]; face from side [Dave Smallshire]; flight and hanging, both [Theo Douma (Agami.nl)]. **p143: Greater and Lesser Horseshoe Bats**, both [Theo Douma (Agami.nl)]. **p144: Lesser Horseshoe Bat** (face) [Andy and Gill Swash (WorldWildlifeImages.com)]. **p145: Lesser Horseshoe Bat**, all [Theo Douma (Agami.nl)]. **p147: Serotine**: flight [Paul van Hoof (Buiten-beeld/Minden Pictures/FLPA)]; at rest [Theo Douma (Agami.nl)]. **p149: Noctule**, both [Theo Douma (Agami.nl)]. **p150: Noctule** [Theo Douma (Agami.nl)]. **p151: Noctule**: at rest and ear, both [Theo Douma (Agami.nl)]; **Leisler's Bat**: at rest and ear, both [Dietmar Nill (Minden Pictures/FLPA)]. **p153: Leisler's Bat** [Franz Christoph Robi (Imagebroker/FLPA)]. **p155: Common Pipistrelle**: flight, both, and wing [Theo Douma (Agami.nl)]; face [Menno van Duijn (Agami.nl)]. **p156: Common Pipistrelle** [Andy and Gill Swash (WorldWildlifeImages.com)]. **p157: Nathusius' Pipistrelle**: at rest [Theo Douma (Agami.nl)]; penis [Daniel Whitby]; **Common Pipistrelle**: at rest [Menno van Duijn (Agami.nl)]; penis [Andy and Gill Swash (WorldWildlifeImages.com)]; **Soprano Pipistrelle**: at rest [Laurie Campbell (lauriecampbell.com)]; penis [Daniel Hargreaves]. **p159: Soprano Pipistrelle**: flight and wing, both [Juniors Bildarchiv GmbH / Alamy Stock Photo]; dentition [Daniel Whitby]; head [Theo Douma (Agami.nl)]. **p161: Nathusius' Pipistrelle**: flight and wing, both [Theo Douma (Agami.nl)]; dentition [Daniel Whitby], illustration [Robert Still]; at rest [Milos Anděra (naturephoto-cz.com)]. **p163: Barbastelle**, both [Theo Douma (Agami.nl)]. **p165: Brown Long-eared Bat**, both [Theo Douma (Agami.nl)]. **p166: Brown Long-eared Bat** (hanging) [Andy and Gill Swash (WorldWildlifeImages.com)]. **p167: Brown Long-eared Bat**: flight and penis, both [Theo Douma (Agami.nl)]; **Grey Long-eared Bat**: flight [Theo Douma (Agami.nl)]; penis (digital artwork) [Robert Still]. **p169: Grey Long-eared Bat**: flight [Juniors Bildarchiv GmbH / Alamy Stock Photo]; at rest [Theo Douma (Agami.nl)]. **p171: Bechstein's Bat**, both [Theo Douma (Agami.nl)]. **p173: Natterer's Bat**: flight [Andy Harmer (flickr.com/photos/38477281@N05)]; at rest [Milos Anděra (naturephoto-cz.com)]. **p175: Daubenton's Bat**, both [Theo Douma (Agami.nl)]. **p176: Daubenton's Bat** [Theo Douma (Agami.nl)]. **p177: Whiskered Bat** [Theo Douma (Agami.nl)]; **Brandt's Bat** [Paul Kennedy]. **p179: Whiskered Bat**: flight [Andy Harmer (flickr.com/photos/38477281@N05)]; dentition [Paul Kennedy]; tragus and penis, both [Daniel Whitby]; at rest [Milos Anděra (naturephoto-cz.com)]; hanging [Theo Douma (Agami.nl)]. **p181: Alcathoe Bat**: flight [Oriol Massana Valeriano and Adrià López-Baucells (adriabaucells.com)]; at rest [Milos Anděra (naturephoto-cz.com)]; dentition and tragus, both [Daniel Whitby]; penis (digital artwork) [Robert Still].

p183: Brandt's Bat: flight [Hugh Clark (FLPA)]; dentition and tragus [Paul Kennedy]; penis [Daniel Whitby]; at rest [Hugh Clark (FLPA)]. **p185: Greater Mouse-eared Bat**, both [Theo Douma (Agami.nl)]. **p186: Northern Bat** [Milos Anděra (naturephoto-cz.com)]. **p187: Particoloured Bat** [Milos Anděra (naturephoto-cz.com)]. **p188: Kuhl's Pipistrelle**: at rest [Michel Rauch (Biosphoto/FLPA)]; dentition and penis, both [Daniel Whitby]; dentition illustration [Robert Still]. **p189: Savi's Pipistrelle** [Milos Anděra]. **p190: European Free-tailed Bat**, both [Leonardo Ancillotto (flickr.com/photos/leonardoancillotto)]. **p191: Hoary Bat** [Michael Durham (Minden Pictures/FLPA)]. **p192: Geoffroy's Bat**: at rest [Milos Anděra (naturephoto-cz.com)]; head [Daniel Whitby]. **p193: Pond Bat** [Theo Douma (Agami.nl)]. **p194: Big Brown Bat** [Fyn Kynd Photography/CC]. **p195: Silver-haired Bat** [Jason Butler (flickr.com/photos/jason_butler)]; **Little Brown Bat** [James Baughman (flickr.com/photos/jblorx)].

CARNIVORES p196: Wildcat [James Lowen (pbase.com/james_lowen)]; **Red Fox** [Mike Hazzledine (flickr.com/photos/64130468@N02)]; **Badger** [Greg and Yvonne Dean (WorldWildlifeImages.com)]. **p197: Pine Marten** [James Lowen (pbase.com/james_lowen)]; **Polecat** [David Kjaer (davidkjaer.com)]; **Stoat** [Peter G Trimming/CC]; **Weasel** [Phil Winter (flickr.com/photos/philwinter)]; **American Mink** [Greg and Yvonne Dean (WorldWildlifeImages.com)]; **Otter** [Hugh Harrop (shetlandwildlife.co.uk)]. **p198: Wildcat** [Laurie Campbell (lauriecampbell.com)]. **p199: Wildcats** [David Kjaer (davidkjaer.com)]. **p200: Wildcat x Feral Cat hybrid** [Sebastian Kennerknecht (Minden Pictures/FLPA)]. **p201: Feral Cat** [Andy and Gill Swash (WorldWildlifeImages.com)]. **p202: Fox** [Pete Adey]. **p203: Fox** [Colin Varndell (colinvarndell.co.uk)]. **p204: Fox** [Kristin Wilmers (Agami.nl)]. **p205: Fox**: cubs play-fighting [Phil Winter (flickr.com/photos/philwinter)]; adult and cubs at earth [David Kjaer (davidkjaer.com)]. **p207: Badger**: pair [David Kjaer (davidkjaer.com)]; single individual from side [Peter G Trimming/CC]. **p209: Badger**: run [Andy and Gill Swash (WorldWildlifeImages.com)]; in sett entrance [Peter G Trimming/CC]. **p211: Otter**: standing [Roger Tidman]; feeding [Andy Oldacre]. **p212: Otters** [Martin Bennett]. **p213: Otter** [Mike Read (mikeread.co.uk)]. **p214: American Mink** [Mike Read (mikeread.co.uk)]. **p215: American Mink**: head [David Kjaer (davidkjaer.com)]; swimming [Colin Varndell (colinvarndell.co.uk)]. **p216: Pine Marten** [David Kjaer (davidkjaer.com)]. **p217: Pine Marten** [Mike Read (mikeread.co.uk)]. **p218: Polecat** [David Kjaer (davidkjaer.com)]. **p219: Ferret** [Karen Bullock (flickr.com/photos/karen_cb)]. **p221: Stoat**: summer [Philip Hunton (Shutterstock)]; winter [Chris van Rijswijk (Agami.nl)]. **p222: Stoat** [David Kjaer (davidkjaer.com)]. **p223: Stoat** [Greg and Yvonne Dean (WorldWildlifeImages.com)]. **p224: Weasel** [Phil Winter (flickr.com/photos/philwinter)]. **p225: Weasel** with prey [Mark Hows (flickr.com/photos/markhows)].

WILD BOAR, DEER, GOAT, SHEEP, CATTLE, HORSE AND WALLABY pp226 & 227: Wild Boar, both [Mark Hows (flickr.com/photos/markhows)]. **p228: Red Deer**: stag [Mike Read (mikeread.co.uk)]; rear [Martin Jones Photography (treelinephotography.co.uk)]; **Sika**: stag [David Kjaer (davidkjaer.com)]; rear [Colin Varndell (colinvarndell.co.uk)]. **p229: Chinese Water Deer**: doe [James Lowen (pbase.com/james_lowen)]; rear [Erni (Shutterstock)]; **Reeves' Muntjac**: buck [James Lowen (pbase.com/james_lowen)]; rear [Colin Varndell (colinvarndell.co.uk)]; **European Roe Deer**: buck and rear, both [Phil Winter (flickr.com/photos/philwinter)]; doe rear [Mike Read (mikeread.co.uk)]; **Fallow Deer**: buck [James Lowen (pbase.com/james_lowen)]; rear [Martin Jones Photography (treelinephotography.co.uk)]. **p231: European Roe Deer**: buck and doe, both [Phil Winter (flickr.com/photos/philwinter)]. **p232: European Roe Deer**: fawn, lying [James Lowen (pbase.com/james_lowen)]; fawn, standing [Colin Varndell (colinvarndell.co.uk)]; does running [Pete Adey]. **p233: European Roe Deer**: running [David Kjaer (davidkjaer.com)]; buck and doe [Phil Winter (flickr.com/photos/philwinter)]. **p235: Red Deer**: stag [Hugh Harrop (shetlandwildlife.co.uk)]; hind and calf [Colin Varndell (colinvarndell.co.uk)]. **p237: Red Deer**: stag [Erni (Shutterstock)]; stag, hind and calf [Mike Read (mikeread.co.uk)]. **p239: Sika**: stag [David Kjaer (davidkjaer.com)]; hind [James Lowen (pbase.com/james_lowen)]. **p240: Sika** [Colin Varndell (colinvarndell.co.uk)]. **p241: Sika**, both [David Kjaer (davidkjaer.com)]. **p243: Fallow Deer**: buck [Martin Bennett]; doe [Greg and Yvonne Dean (WorldWildlifeImages.com)]. **p244: Fallow Deer**: fawn [Martin Jones Photography (treelinephotography.co.uk)]; buck [Danny Green (Agami.nl)]. **p245: Fallow Deer**: does, both [James Lowen (pbase.com/james_lowen)]. **p247: Reeves' Muntjac**: doe [Mark Hows (flickr.com/photos/markhows)]; buck [Colin Varndell (colinvarndell.co.uk)]. **p249: Chinese Water Deer**: doe [James Lowen (pbase.com/james_lowen)]; buck [Erni (Shutterstock)]. **p250: Reindeer**: summer [Sergey Gorshkov (Agami.nl)]; winter [Chris van Rijswijk (Agami.nl)]. **p251:** Reindeer [Sergey Gorshkov (Agami.nl)]. **p253: Feral Goat**, both [Mike Read (mikeread.co.uk)]. **p255: Soay Sheep** [Colin Monteath (Minden Pictures/FLPA)]; **Boreray ram** [John Eveson (FLPA)]. **p256: Chillingham Cattle** herd [David Clay/CC]. **p257: Chillingham Cattle**: two animals [John Lord/CC]; bull [Thomas Quine/CC]. **p259: Dartmoor Pony** [Mike Read (mikeread.co.uk)]; **Exmoor Pony** [Mark Hows (flickr.com/photos/markhows)]. **p260: New Forest Ponies** [David Kjaer (davidkjaer.com)]; **Welsh Mountain Pony** [Mark Hows (flickr.com/photos/markhows)]. **p261: Red-necked Wallaby** [Mike Read (mikeread.co.uk)].

Finally, the authors would like to express their own personal debts of gratitude.

Dominic Couzens would like to thank his wife Carolyn and children Emmie and Sam for their patience and understanding during the long and often torturous process of putting this book together. Thank you for your love and support over the last three years.

Andy Swash is, as ever, indebted to his wife Gill for her unstinting support during the production of this book, and for her insightful comments on the text during the final stages of production. He would also like to thank Mum for her constant encouragement.

Rob Still thanks The Pixies, the late Lou Reed and the late David Bowie for the primary soundtrack to the completion of this volume, Rachel Emily for her valued creative and organisational input, Anya Naomi for the supply of quesadillas, and all who have helped bring joy into my life.

Jon Dunn would like to thank fiancée Roberta for her love and support – she is the cat's vibrissae – and son Ethan for being cooler than a Narwhal's shades. Particular thanks are due to his co-authors for all their hard work and dedication.

Index

This index includes the English and *scientific* (in *italics*) names of all the mammals in this book.

Bold text highlights main species accounts.
Brown figures indicate keys or comparison spreads/tables.
Green figures indicate illustrations of tracks and signs.
Italicized figures indicate page(s) on which other photographs appear.
Blue figures relate to the entry in the status and legislation table.
Regular text is used for alternative names, species that are not subject to a full account, and to indicate pages where other information may be found.

About the authors

Dominic Couzens is one of Britain's best-known wildlife writers. His work appears in numerous magazines, including *BBC Wildlife* and *BBC Countryfile*, and his books include *Secret Lives of Garden Wildlife* and *Save our Species*.

Andy Swash is an ecologist, a renowned wildlife photographer, and managing director of WILD*Guides*. He has co-authored seven other books in the **WILD***Guides Britain's Wildlife* and *Europe's Wildlife* series, and written and edited many other titles.

Robert Still, co-founder and publishing director of **WILD***Guides*, is an ecologist and widely travelled naturalist. His design philosophy and exceptional skills in computer graphics have been key to the development and production of the **WILD***Guides Britain's Wildlife* and *Europe's Wildlife* series.

Jon Dunn is a natural history writer, photographer and wildlife tour leader based in the Shetland Isles, UK. His books include *Orchid Summer* and *The Glitter in the Green*.